Science Reforged:

The New Mathematics of Epistemology

&

Why the Soft Sciences are Soft

Richard Howard Kramer

New Haven, Connecticut

For Karl Popper and Ian Hacking,
who almost got it right.

Abstract.

> The logic now in use serves rather to fix and give stability to the errors which have their foundations in commonly received notions than to help the search after truth. So it does more harm than good.
>
> —Novum Organum Scientiarum, Liber Primus, XII

Given a finite set of hypotheses that *might* be true and given that one of them *is* true, a method is presented for choosing among them based on experimental evidence with the equal probability of choosing correctly no matter which hypothesis is true. This provides a firm foundation for the hard sciences.

With an additional postulate, the method can be extended to a finite set of families of hypotheses. An example of a family of hypotheses would be all normal distributions with mean zero. With this additional postulate, the method can be extended to the soft sciences. Unfortunately, applied to the soft sciences, it uncovers a flimsy foundation.

The hard sciences get it right without knowing why; the soft sciences get it wrong without knowing why.

Contents

Preface

There are two kinds of scientist. The first kind has a specific result in mind and tests it with an experiment. When the result is spot on, he is happy. If the result is wide of the mark, he is upset and hopes to find an alternative hypothesis that explains what he thought he knew before, as well as the anomalous result. When testing two alternative hypotheses he feels justified in rejecting one over the other if the experimental results for one are spot on and are five sigmas away from the prediction of the other.

The second kind of scientist reports his results in terms of statistical significance or confidence intervals. Before the experiment, he has only vague directional expectations. He has no specific numerical target in mind. It is the difference between explaining gravity by $F = \frac{Gm_1m_2}{r^2}$ or "Apples fall."

The exemplars of the first kind of science are physics, chemistry, and other of the physical sciences. Sciences of the second kind include psychology, sociology, economics, epidemiology, medicine, and other soft sciences.

The difference in results is only obliquely related to the subject matter, but rather to the *form* of the hypotheses in question. The decisiveness of a test depends critically on the existence or nonexistence of clear, specific hypotheses on the one hand and vague, hard to quantify hypotheses on the other.

It is the intention scientists of the second type to produce results that are in some way comparable to science of the first type. By presenting their results in terms of statistics, scientists of the second type believe they are doing something like what scientists of the first type are doing. They suffer from *physics envy.*

Scientists of the second type are not really making their results more scientific by the use of the statistical methods they learned in graduate school. The reason is that those techniques do not directly address the probable truth of alternative hypotheses; the techniques themselves are flawed.

The numerical results of significance testing and the probable truth of alternative hypotheses are orders of magnitude different. A great many of the results that are assumed to have been justified by experimental data are so weakly justified that they should not be relied on. In fact, it is often the case that the results dissolve into nothingness upon application outside the original data set.

It has not escaped notice that significance testing and confidence intervals have produced many flawed results:

"Mindless Statistics" [**16**],
"Why Most Published Research Findings Are False" [**24**],
"How Persuasive Is a Good Fit? A Comment on Theory Testing" [**41**],
"The Fallacy of the Null Hypothesis in Soft Psychology" [**51**],
"Consequences of Prejudice Against the Null Hypothesis" [**17**].

The realization has even moved into the popular press,

"The Truth Wears Off" [**30**].

"Repeat Performance: Too Many Studies, When Replicated, Fail To Pass Muster" [**43**]

Everyone knows that economic forecasts are not very good, and that coffee is good for you, bad for you, good for you. Everyone knows that lots and lots of clinical trials produce results that are not durable. Everyone knows that sociological facts when put into practice to change society tend not to work so well. Why do these seemingly "well-tested and well-justified hypotheses" fail to produce reliable, durable results?

It will be shown that many of the results produced by scientists of the second type are *not justified.*

Imagine a soft scientist playing Russian roulette with a five-shot revolver loaded by someone else. The revolver may contain zero to five bullets. He spins the cylinder, pulls the trigger and survives. He then announces, "Based on this experiment, there is *one chance in one thousand* that this revolver contains *any* bullets."

If the "experiment" had been done with a six-shooter, the odds would have been in the neighborhood of *one chance in twenty thousand.*

This is how much the soft sciences overvalue their results.

Part I

Fables and Examples

CHAPTER 1

Introduction to Part I

Part I consists of three of the previously lost stories of Sindbad the Sailor. In these tales, Sindbad is required to make judgments about what would now be called statistical inference. He does so in a spirit quite foreign to current practice, but his method is so simple and intuitive, it is hoped that none can doubt the method.

There is some indication that the manuscripts in question once resided in the Bayt alHikma, The House of Wisdom. The text is in Arabic, but there are annotations in Chinese concerning the mathematical justification of Sindbad's conclusions. This makes it likely that these texts were among the manuscripts removed from Baghdad by Nasīr alDīn Tūsī before the sacking of the city by Hulagu Khan. It is quite possible that the reason that these three tales remained undiscovered for so long is that they traveled east along the Silk Road.[1]

[1]Or perhaps, they are pseudepigraphy.

CHAPTER 2

Sindbad in the Cave of Gold and Silver

QUOTH SHAHRAZAD:

Then Sindbad took the turbans from the heads of the guards he had slain and tied them, end to end, to make a rope. He tied one end of the rope around the base of the column supporting the railing of the balcony on the side opposite from where the great elm grew in the garden beside the palace wall. He gave the princess one last kiss and then, as the rush of footsteps and shouts of alarm sounded in his ears, he stretched the rope of turbans across the balcony to the side nearer the tree and lowered himself hand over hand from her bed chamber.

Halfway to the ground, by pushing against the wall, Sindbad was able to swing into the upper branches of the elm. Then with a quick flip of his wrist, Sindbad dislodged the rope so that it hung to the ground from the *far* side of the balcony more than thirty feet away from where he hid.

Soon soldiers were rushing about. They would look at the rope and look at the balcony. Then they would run off in some direction, yelling orders or cursing. Never did they look into the branches of the tree.

Sindbad knew, however, that no hiding place could hide him from hunger. How could he escape from the outer walls? Could he disguise himself and walk boldly out the gate? Would he be recognized? Where could he find a disguise?

As these thoughts vexed Sindbad's mind, his eyes grew weary and he drifted off to sleep. A dream from his childhood came to him. His grandmother was telling of events long ago when raiders came from the wide grasslands to the north. They had come before his grandmother was born, burning the fields, stealing the grain and cattle, killing, and raping. Treachery had opened the gates to the capital and most had perished, but the Shah had escaped by climbing down a well that led to a secret tunnel under the walls. He was able to steal away, raise a second army, and, in the end, defeat the invaders and restore the dynasty.

Sindbad awoke with a start! Could it be that the dream was true? Had his grandmother told this tale? Was the tale itself true? When he tried to recover a memory separate from the dream, he could not. When he tried to remember more details of the story of the escape through the well, he could remember none.

He looked out across the palace grounds. In a distant corner there *was* a well that seemed unused. No path led to it; it was overgrown with vines and flowers. No bucket sat on its edge, but there was a rope hanging down into it. Could this be his path to freedom?

Sindbad knew if he ventured from the tree, he could never return. There was no way to climb back up, no low branch. If he went to the well and there was no way of escape, he would almost surely be discovered and killed.

In a flash, the decision was made! Sindbad hung from the lowest branch and dropped to the ground. Catlike, he crouched in the shadows and moved noiselessly across the grounds. Several times guards came so close to him that, but for his skill and God's favor, he would have been found out.

At last he reached the well. Indeed, it had not been used in scores of years. Around the top there was an intricate design that had worked into it words in a language unknown to Sindbad. Much was written, but Sindbad could make out only this: "*Lasciate ogni speranza, voi ch' entrate.*"[1] The rope, though sound when new, seemed not altogether trustworthy now, but he had no choice. There were footsteps and voices behind him. Over the edge and onto the rope he went. Hand over hand, he let himself down, straining to see the bottom in the darkness. Then his imprudence made itself known. He reached the frayed end of the rope! Still he could see or feel no sign of the bottom of the well.

Sindbad forced himself think. If this was a secret escape route from the palace, the rope would not go all the way to the bottom lest the outside entrance be discovered and the passage used to enter the palace. On the other hand it might just be an old well, gone dry. For a second time in one night, Sindbad trusted his fate—he dropped into the darkness of the well.

He seemed to drop forever. Gradually a faint light appeared, as if a torch were burning somewhere below. Faster and faster he fell, until, almost as in a dream, he dropped into an enormous cushion, so soft it must have been filled with the down of ten thousand geese.

A massive, but strangely shaped, attendant greeted him thus: "Master Sindbad, welcome to the caves of fate. Some, not many, have left here wealthy men; most have simply ended as my dinner."

Sindbad drew his sword and crouched to repel the anticipated attack, but the strange giant just smiled. "You had better use your wits, rather than your sword. With them you have some chance of victory," said the demon, drawing a sword so large and unfamiliar that it seemed from another world. The sword glistened and moved like oil on water. Near the tip, its name was written: "*Rozar S'macco.*"

Thinking better of his position, Sindbad returned his sword to its sheath. "How am I to be tested?"

"Ah, Master Sindbad, you must enter three caves and answer three riddles. Each time you answer correctly, you will be given the riches of the cave. If you answer all three riddles, I must set you free. If you fail, . . . well, we will not talk of that now.

"When you are ready, you may enter the Cave of Gold and Silver," said the giant pointing to a low arched entrance ornamented in the Greek style with glass tiles backed by precious metals.

Sindbad bent low and entered the passageway. It was perhaps twenty paces long, becoming taller and narrower as it approached the cave itself. As he neared the exit, Sindbad could see the light of many torches playing on the walls. These walls, too, were covered with gold and silver tiles. The light sparkled like sunlight on water, so bright the water could not be seen, only flashes of light.

To enter the room before him, Sindbad had to turn sideways and squeeze through the opening. As he did this, there occurred a wonder that causes me to tremble at the telling lest you, O my King, disbelieve all that follows. The demon by some means that I know not, had already entered the cave and was holding his sword high above his head. He brought it down with the sound of thunder, cutting Sindbad in half from top to bottom across his shoulders. He fell to the floor, front half to east and back half to the west.

To Sindbad it seemed that something very different had happened. It seemed to him that the great sword had come down very close to his face but had not touched him. Trailing the sword there was a shimmering reflection like an imperfect mirror. In the mirror was his own reflection, but it seemed only *he* was reflected, for looking past his image, he could still see what he had seen before.

When he moved the image was not always inclined to follow. When he spoke, the image seemed to consider for a moment before it spoke too. At last Sindbad dared to reach out and touch the image. It

[1]These words also appear in a long Italian poem published many centuries after these events [1]. This poem is one of a series of poems written over more than two millennia to elucidate a large compendium of footnotes that has survived from prehistory. This poem was previously believed to be the first of its type. The dating of the present manuscript indicates that the poem itself may have deeper roots.

reached out to him as well, but unlike a mirror, it grasped his hand. Both Sindbad and his image sprang back in astonishment.

At this point, the giant could restrain himself no more. He began to laugh with such force that some loose tiles fell from the ceiling and shattered on the floor. At last after a very long time indeed, he was able to speak and said this: "My good fellow, it is quite true! You were one, now you are two. It's all necessary for your first test.

"As you will soon notice, the two of you will not easily agree on anything, but to avoid a very unpleasant end, you must find a way to force the other to agreement even though *he* will not believe as you do.

"Here is the problem the two of you must solve: In the cave there are small leather bags beyond counting. Each contains five coins. All the bags contain the same configuration of coins. There are two possibilities: either the bags all contain three gold coins and two silver coins or the bags all contain three silver coins and two gold. You are allowed to pick one bag and, without looking, from that bag pick one coin. You must then tell me what all the other bags contain. You must both agree, before drawing the coin from the bag, how you will answer, and you must be able to justify your answer."

At this point, Sindbad's mind was trying frantically to guess what traps the demon had laid. As he thought, he could hear his reflection thinking, but their thoughts were always moving in opposite directions. When he followed a line of reasoning that led to *silver*, his reflection would follow the same argument and come to *gold*.

At last, exhausted, Sindbad and his image abandoned trying to out-think the demon and decided to trust to luck. Even then it was not easy to come to agreement. At last they picked a bag and, as if one, picked a coin from the bag and without looking at it, said in unison, "The other bags contain three coins like this one and two coins unlike it."

Clearly, the demon was displeased. "That is only the first part of the task. You must also justify your answer."

Panic welled up in Sindbad's heart for he could see that although he thought the coin they held in their hand was most likely *gold*, his double thought it was most likely *silver*. Sindbad's calculations went like this: "I believe that there are three gold coins and two silver coins in the bags. Therefore there are three chances in five that this coin is gold and correctly reflects the odds of this bag and the rest of the bags."

At the same time, Sindbad's image was thinking the opposite: "I believe that the bags contain three silver coins and two gold ones. Therefore there are three chances in five that this coin is silver and correctly reflects the contents of this bag and the others in the cave."

In order to clear their thinking, Sindbad and his image drew à table of possibilities on the floor of the cave.

	The Coin Drawn from the Bag:	
	gold coin	*silver coin*
3 *gold coins* 2 *silver coins*	$\frac{3}{5}$	
2 *gold coins* 3 *silver coins*		$\frac{3}{5}$
	say: "most of the coins are *gold coins*"	say: "most of the coins are *silver coins*"
	The Decision	

TABLE 2.1. The Cave of Gold & Silver

Each row represented a possible arrangement of coins in the bag. The left column represented drawing a gold coin from the bag; the right column represented drawing a silver coin. The value in each cell represented the chances of correctly drawing a particular kind of coin, given the arrangement of coins represented by that row. For instance, in the row labeled "Two gold coins and three silver coins,"three chances in five of drawing a silver coin. At the bottom of each column, is the pronouncement to be made if the drawn coin matches that column.

In the end, Sindbad spoke: "Even though each of us believes that the coin that was picked from a different kind of bag, we each find it possible to contents that there were three chances in five that, that coin was the same kind as the majority we each separately believe to be in all the bags.

"The decision rule we have chosen is the best one because it is the one decision rule we both can agree on that has the best chance of being correct."[2]

Was it gold or silver? This can not be told, for when the demon heard their answer, he tore the coin from their hands and threw it into the darkness beyond the light of the torches.

"Thus ends the tale of the Cave of Gold and Silver," said Shahrazad, as the rosy fingers of dawn reached across the lake to the pavilion where Shahrayar and Shahrazad reclined on cushions, sipping tea flavored with cardamom and mint.

Shahrayar kept his own counsel for a long time and finally spoke. "How can I know that Sindbad's answer is the best? He chose to say that most coins in the bags were like the coin he picked, but there are three other decision rules he could have followed. He could have said the bags contained the *opposite* of the coin picked from the bag or he could have said 'gold' no matter what he drew from the bag or, finally, he could have said 'silver,' irrespective of what was drawn from the bag."

Shahrazad replied, "O, my King, if he said the *opposite*, whether he believed the bags had three gold coins or three silver, he would have only two chances in five to be correct. The other two decision rules could not lead to agreement, for either Sindbad or his image would have preferred the decision rule they chose to one that always said 'silver,' or that always said 'gold.' Further tests await Sindbad more intricate than this, but they must remain untold until tomorrow."

[2]This result was derived independently by Ian Hacking many centuries later [**18**, chapter IX]. Although his derivation is different, Hacking saw that, in this special case, the probability of the evidence, given rival hypotheses, is transferred intact to probability the truth of the rival hypotheses, given the evidence.

CHAPTER 3

Sindbad in the Cave of Rubies and Sapphires

Shahrayar and Shahrazad supped on lamb steeped in the juice of pomegranates and honey; lentils with spinach, garlic, and lemon; rice with chicken, almonds, and dates; and a hundred other things. Finally, they sat looking at the moon, eating sherbet made of the beaten flesh of fruit, and of snow carried by special courier from the mountains.

Shahrayar spoke first. "Tell me of the trials of Sindbad and the demon who held him captive."

QUOTH SHAHRAZAD:

> Though he was evil, the demon was bound by his word. He paced back and forth in the cave waving his arms and sometimes shouting at Sindbad and his image, and sometimes muttering to himself. In the end he turned to them and said, "There is no denying it, you solved the riddle correctly. Your reward will be one hundred bags of coins on each of one hundred camels.
>
> "There is one condition, as I mentioned before. Before you can claim your reward, you must solve the riddles of the remaining two caves." With this he motioned to a flight of narrow steps that led to a small doorway near the ceiling. Around the door was a frame of woven and twisted gold wire much like the adornments worn in Granada. The door, itself, was covered with mosaics made of tiny sapphires and rubies. The demon stepped forward to open the door. Beyond it was a smooth spiraling passage that could only be entered by sliding on one's back. Quick as the blink of an eye, the demon pushed one of the Sindbads into the opening and seized the other by the neck. Out came his sword; as a street vendor slices roast lamb, so the demon cut the second Sindbad into three pieces and tossed them into the spiral passage. Then he slammed the door and disappeared.
>
> As the twice and thrice chopped Sindbad tumbled down the chute, he heard a low sound resonating around him. As he descended, the sound twice rose in pitch as a vibrating wire divides in half and then thirds as it is stretched tauter and tauter. When they spilled out on the floor, it was not a jumble of bloody parts that emerged, but one large and three tiny Sindbads, each speaking at cross purposes to the others.
>
> The demon was already there waiting, arms crossed, beside the larger Sindbad. When the full-sized Sindbad saw his three small counterparts, he rushed forward to touch them for he thought they might only be a dream, but they were as real as he or any man.
>
> "Silence," roared the demon. "If any of you are to leave here alive, you must solve the riddle of the Cave of Rubies and Sapphires. Here, as in the first cave, there are leather bags without number. Each bag contains five gems the size of a quail eggs, a combination of sapphires and rubies. All the bags are alike in their contents. There are only four possibilities allowed: the bags contain five rubies and no sapphires, four rubies and one sapphire, three rubies and two sapphires, or two rubies and three sapphires. Your task is to remove one gem from one bag and guess whether there are more rubies than sapphires in the bags. As before, you all must agree and must justify your answer."
>
> If your retainers cast golden coins at the feet of one hundred beggars, *that* O my King, is the only way to describe what followed. The larger Sindbad and the three small Sindbads yelled and shoved and yelled once again. Each had a different opinion about the contents of the bags. When they agreed as

to method, they disagreed about the values that should be used to implement the method. Finally they agreed, but only on a procedure for discussion.

The larger Sindbad agreed to represent the belief that the bags contained three sapphires and the three smaller Sindbads, that the bags contained two, one, or none. Thus between them, they covered all the possibilities and the division between the larger Sindbad and the committee of smaller Sindbads represented the decision that they were called upon to make. Contentious as they were, they agreed to this because no matter what they actually believed to be true, it was certain to be represented by one of them.

Then the discussions went more quietly. Soon they found a decision rule with which all could agree. Not long after, they found that there *did* exist a way to justify their decision. To help explain their strategy the large Sindbad drew a table on the floor of the cave listing the possibilities.

	The Gem Drawn from the Bag:	
	ruby	*sapphire*
5 *rubies* 0 *sapphires*	1	
4 *rubies* 1 *sapphire*	$\frac{4}{5}$	
3 *rubies* 2 *sapphires*	$\frac{3}{5}$	
2 *rubies* 3 *sapphires*		$\frac{3}{5}$
	say: "most of the gems are *rubies*"	say: "most of the gems are *sapphires*"
	The Decision	

TABLE 3.1. The Cave of Rubies & Sapphires

The larger Sindbad reached in to a bag, withdrew a gem and said, "There are three or *more* chances in five that bags contain a majority of gems of the same kind as the one I hold in my hand."

"Prove it, prove it," screamed the demon. "Your opinion means nothing to me—I require facts, I require proof."

"Our opinions enter into the decision, not at all," said the larger Sindbad. "The proof is to consider each of the possibilities in turn. In each case, the statement I have made is true. We have relied only on what you said to be true and the very meaning of chance. If the bags contain three rubies and two sapphires, or three sapphires and two rubies, the problem is the same as the riddle of the Cave of Gold and Silver—the answer is three chances in five, exactly. If there are four rubies and one sapphire in the bags, there are four chances in five of drawing a ruby. Thus there are four chances in five that my statement is true—four chances in five is surely greater or equal to three chances in five. The final possibility, five rubies and no sapphires, follows the same reasoning."

The demon drew his sword and began to chase the four Sindbads around the cave. Though he was fast, fear outran anger. The demon did not catch them and, finally, began to hack at the wall instead. After a quarter of an hour, his fury exhausted, he slumped against a pillar and sighed.

"You are correct. You shall have ten thousand slave-girls, each carrying ten bags of sapphires and rubies as your reward. But first you must solve the riddle of the third cave."

With this Shahrazad yawned and looked to the east where the first morning light was chasing the stars from the sky. "O my King, tonight I will continue the story of Sindbad in the demon's caves, but now I am too weary."

Thus ends the tale of Sindbad in the Cave of Rubies and Sapphires.

CHAPTER 4

Sindbad in the Cave of Diamonds and Emeralds

The next day in the afternoon Shahrayar greeted ambassadors from the great kingdom to the east. They rode horses, as large and beautiful as any Shahrayar had ever seen, not at all like the ponies used by the herdsmen to the north. Their clothing was of fine glistening fabrics covered with strange designs. When they ate they had servants cut their food into small pieces and then used sticks to put the food into their mouths rather than eating in the ordinary way.

When he retired to his chambers, Shahrayar was weary of diplomacy, rumours of war, exchange of gifts, disputes with the towns along the trade route beside the great desert—all too much to bear until another day.

As he entered, Shahrazad put down the book she was reading. "Come here. Lie by me," she said. She dipped a cloth into the rose scented water that flowed from a fountain in the wall into a basin below and wiped his face.

"I am weary of the world's dealings," he said. "Tell me of the adventures of Sindbad in the third cave."

QUOTH SHAHRAZAD:

> Since there seemed to be no escape, Sindbad and his three tiny brothers saw no choice but to follow the demon as he walked into a small side cave and began to climb a ladder formed of brass spikes driven into the wall. The rungs of the ladder were too far apart for his three small companions to climb, so they held on to his clothes. Sindbad climbed at least one hundred steps up the wall of the cavern. At the top, they stood by a pit whose sides sloped downwards and toward the center until they ended in an opening the size of a man's shoulders. The surface of the sloping sides shimmered as if it were made of quicksilver and spiraled downward, alternating green and silver stripes—like a whirlwind must look when seen from the inside of the cloud from which it descends.
>
> "This is the opening to the Cave of Diamonds and Emeralds," said the demon. "I will meet you within the cave," he said, pushing Sindbad into the pit, and then the others after him with his great arm. They struggled to keep their footing, but could not, and one by one, slid down the sloping side and into the hole in the center of the pit. They fell for a long time, until they eventually began to hear a high pitched whirring noise, not unlike the wings of a hummingbird, but louder and with a metallic ring.
>
> In the chamber below, the demon was slowly turning the crank handle of a great machine, gear upon gear, multiplied the revolutions of his turning until finally his slow motion was transferred into the motion of an array of swords, moving so fast they could not be seen except as a blurring. They were similar to those Sargon the Great put on the wheels of his chariots to strike terror into the hearts of his enemies. Above the spinning swords was a massive funnel.
>
> Sindbad and his compatriots fell through the blades. They were cut into pieces as numberless as the sands, but rather than simply being cut into gobbets, each piece was a minuscule Sindbad, perfect in every way. There were two groups of Sindbads. One group was formed from the original Sindbad and the other was formed from his thrice-divided image. All the tiny Sindbads were talking and yelling at the same time, but in the confusion, their voices merged into a sound like the ocean when it runs up against the shore.
>
> The demon smiled. Holding up a leather bag he said, "This, my small friends, is the riddle of the Cave of Diamonds and Emeralds: this bag is filled with Diamonds and Emeralds. There are two possibilities.

The bag may contain two diamonds and at least three emeralds or it may contain two emeralds and at least three diamonds. In both cases, there are only two of the jewels that are in the minority, but there may be numberless ones of the jewel in the majority.

"Your task is to remove one jewel from the bag, then tell me whether the bag contains mostly diamonds or mostly emeralds. That is the easy part. In addition, you must justify your choice unanimously."

Now the little Sindbads and the little images of Sindbad had some experience with this kind of riddle before, and thus had an idea of what to do. Each of them was assigned a possible contents of the jewels in the bag. The little Sindbads took the cases in which most of the gems were diamonds; the little images of the three Sindbads took the cases in which most of the jewels were emeralds. They enlisted the aid of the demon, for none of them, now, was large enough to hold a *qalam*. On their instructions, he created a table of possibilities on the floor of the cave.

	The Gem Drawn from the Bag:	
	diamond	*emerald*
$3{+}n$ *diamonds* 2 *emeralds*	$\frac{3+n}{5+n}$	
⋮	⋮	
4 *diamonds* 2 *emeralds*	$\frac{4}{6}$	
3 *diamonds* 2 *emeralds*	$\frac{3}{5}$	
2 *diamonds* 3 *emeralds*		$\frac{3}{5}$
2 *diamonds* 4 *emeralds*		$\frac{4}{6}$
⋮		⋮
2 *diamonds* $3{+}n$ *emeralds*		$\frac{3+n}{5+n}$
	say: "most of the gems are *diamonds*"	say: "most of the gems are *emeralds*"
	The Decision	

TABLE 4.1. The Cave of Diamonds & Emeralds

In what seemed like an endless amount of time, being so many, they elected a spokesman who went to the demon and said, "To solve the riddle of the cave of Diamonds and Emeralds, one should withdraw the gem and say, "There are at *least* three chances in five that most of the jewels in the bag are of the same kind as the one I hold in my hand." The justification is the same as before! Every other decision rule has a smaller chance of being correct as viewed from the position of at least one of us. We can all agree on the statement made, though to all but two of us, there is a *greater* chance of being correct."

Words were spoken from afar; it seemed that the cave was filled with lightning and thunder. The demon glowed red and then yellow and then white. He began to melt and run across the floor of the cave like molten iron. The tiny Sindbads ran for their lives, but there was no escape. When the fiery liquid overtook them, they burned so that all that was left was thick smoke. The smoke rose to the top of the cave and then up through a natural vent in the rock. It rose and rose. As it rose, it seemed to attract itself and form a kind of rope spiraling up and up. Finally it could rise no more and formed into the shape of a man, clutching the rope in his sleep. The man bore a resemblance to Sindbad. He was sitting cross-legged, slumped forward, but he was upside down. He appeared to be sleeping, head down, suspended by the seat of his trousers.

This strange figure opened his eyes, blinked, and let go of the rope. He seemed to float upward toward the ceiling until he arrived with a thump. Instead of the ceiling, he found himself sitting on the floor at the bottom of a well. High above he could hear soldiers running back and forth across the palace grounds.

In his right hand, he held a diamond; in his left, an emerald.

At this point, Shahrazad paused and said, "O my King, I am weary. Tomorrow I will continue the tale of Sindbad and tell how he was able to escape from the well and from the palace grounds."

"Wait! wait!" shouted Shahrayar. "I do not see how the answer to the third riddle is correct."

"O my King, consider, in each row of the table, whether the decision rule proposed gives the correct answer. You will find that the decision rule is correct in the cases that are above the midpoint and to the left or below the midpoint and to the right. In the upper left, there are always three chances in five, or greater, that a diamond will be chosen. When a diamond is chosen, the decision rule prescribes the correct answer—"Mostly diamonds." Similarly, in the lower right, there always three chances in five, or greater, that an emerald will be chosen. When an emerald is chosen, the decision rule prescribes the correct answer—"Mostly emeralds." In each case, there are at *least* three chances in five that most of the jewels will be of the same kind as that drawn from the bag, even though we do not know which kind it is, until we actually look.

Thus ends the tale of Sindbad in the Cave of Diamonds and Emeralds.

Abandon Hope!

If you have followed the riddles of the three caves, you have the knowledge required to invalidate the mathematical methods used by the soft sciences and thus to cast in doubt most of their results. If you wish to preserve the soft sciences in their present position, you must turn back now. You are at the beginning of a slippery slope and will be swept away as surely as the little Sindbads.

Part II

Mathematical Formalism

CHAPTER 5

Introduction to Part II

In this part, the lessons of Part I are formalized and extended into areas where they intersect with the practices of the sciences, both physical, biological/medical, and social/psychological. An exhaustive catalog of permutations will be examined—all involving the choice between probabilistic hypotheses based on sample evidence represented by normal distributions.

The same techniques could be applied to other distributions, but for simplicity and brevity attention is restricted to the simple case. Almost all definitive tests involve a sample large enough that a normal distribution is an adequate approximation.

For several centuries, philosophers have sought a way to calculate the absolute probability that a hypothesis is true, $\mathrm{P}(\mathfrak{h})$, all without notable success. Often the effort begins with a theory of rationality which in turn eventually leads to a formula that provides for the accretion of positive instances, congruent with the hypothesis, to lead by means of some formula to the probability that the hypothesis is true.

Sometimes the starting point is a stripped-down theory of language, which eventually leads to a theory of probability as in Carnap [**5**].

Sometimes the effort begins by putting hypotheses and evidence statements in the same probability distribution, conjuring up prior probabilities for the hypotheses, either by personal preference or some standard of neutrality. Then Bayes' theorem is used to produce probabilities of each the hypotheses given the evidence.

Only a last of these efforts is active in the present day. The Bayesian strategy has produced many clever ways of estimating parameters, but it is unable to produce the probability of a hypothesis being true unless the prior probabilities are not just plausible, but are in fact true. Flaws in the Bayesian analysis will be addressed in Chapter 26.

In the meantime, practitioners of the soft sciences have adopted the use of significance tests and confidence intervals to report their results. Neither of these strategies directly addresses the question of whether a hypothesis is true.

The *standard procedure* is first to attempt to splice *modus tollens*[1] onto probability. It goes like this: "P implies a very unlikely Q, therefore probably not P, therefore probably Q." The chain of reasoning is absurd, but it is tacitly endorsed by soft scientific practice.

Suppose a sample is on the positive side of zero, the null hypothesis, and a t-test is significant at the $p = .01$ level. To what degree is one justified in believing that reality is like the sample? To what degree is one justified in believing the null hypothesis is true? The t-test gives no answer to either of these questions, yet soft scientists *act* as if it gives *some* answer.

The second way of reporting results in the soft sciences is a confidence interval, usually 95%, constructed from the sample. This is an attempt to show the practical import of the hypothesis represented by the sample, if it were true. The process is just a reframing of a two-tailed significance test at the $\alpha = .05$ level and estimates neither the probability that the null hypothesis is true nor the probability that the hypothesis represented by the sample is true.

Things are quite different in the hard sciences. It is very very seldom that one will see any of these techniques used. This is because, in the hard sciences, the consensus is that these techniques don't work and are irrelevant.

---

[1] *Modus tollendo tollens*, Latin for "the way that denies by denying." P implies Q, but not Q, therefore not P.

The hard sciences have given up on the idea that by the accretion of confirming evidence, one can make the statement that a particular theory is probably true. This strategy just doesn't work—witness various scientific revolutions: Ptolemy to Copernicus, Newton to Einstein, common sense reality to quantum mechanics. Things get interesting when there is a surprise!

The counterexamples are what has causes the hard sciences to embrace Karl Popper's paradigm that a decisive refutation of a theory weeds out error—the last man standing theory of the evolution of knowledge. The refutation is never based on sophisticated statistical methodology; it is a experiment that is five or six standard deviations away from the predictions of the theory in question. The probability distribution of the measurements is determined by the instrumentation. If there is no new theory, there is a problem to be worked out. At the point of a paradigm change, there is a alternative theory and the differences between the two suggested the experiment. If the alternative is spot-on and is five standard deviations from its competitor, it will have scored a knockout blow and be (temporarily) champion.

Unfortunately, is the soft sciences such knockout blows almost never happen because the "theory" is so vague that its predictions are something like "over there somewhere." The fate of most theories in the soft sciences is to fade into irrelevancy or go, untested by replication, to live a comfortable life in textbooks.

In economics, there are a variety of mutually contradictory theories about how the economy actually works. So far as I know, no one has proposed a test that would eliminate one or another of them. They just drift in an out of fashion depending on recent events. Some disappear because their ardent supporters retire from the field because of old age and no one takes up the mantle. Econometric models have negligible accuracy outside of the data used for their estimation, so they provide little guidance.

In psychology and sociology knowledge fades with time and ceases to be effective when it is acted on. There is wide discomfort, but little unanimity as to what is wrong or what to do about it.

The answer is contained in the next chapters. It is a procedure that yields a probability statement about the truth of theories that is *deducible* from the meaning of the theories and the evidence used to test them. It shows why the form of the theories in the soft sciences precludes strong conclusions and why results are suspect.

CHAPTER 6

Definitions

Logical Notation

$\exists a$	means "there exists an a such that..."
$\forall a$	means "for all a..."
$\neg a$	means "not a"
$a \;\&\; b$	means "a and b"
$a \Rightarrow b$	means "if a, then b"
$a \not\Rightarrow b$	means "It is not the case that $a \Rightarrow b$"
$a \Leftrightarrow b$	means "a if and only if b"
$a \not\Leftrightarrow b$	means "It is not the case that $a \Leftrightarrow b$"

Set Notation

$S = \{a, b, c\}$	means "S is a set containing as elements $a, b,$ and c"
$a \in S$	means "a is an element of set S"
$E_{\mathfrak{h}_i} \subseteq E$	means "$E_{\mathfrak{h}_i}$ is a subset of E" $\Leftrightarrow \forall e, e \in E_{\mathfrak{h}_i} \Rightarrow e \in E$

Mathematical Notation

$a = b$	means "a is *numerically* equal to S"
$a \neq b$	means "a is not *numerically* equal to b"
$a \approx b$	means "a is approximately *numerically* equal to b"
$a \propto b$	means "a varies in proportion to b"
$\|a\|$	means "the value of a without regard to sign"
$\min(x_1, x_2, \ldots, x_n)$	means the "minimum of $x_1, x_2, \ldots, x_n$"
$\max(x_1, x_2, \ldots, x_n)$	means the "maximum of $x_1, x_2, \ldots, x_n$"

The usage of $\mathfrak{h} \in \mathfrak{H}$, $e \in E$, and $\ulcorner \boldsymbol{d}(e) \urcorner$

- Let $\mathfrak{H} = \{\mathfrak{h}_1, \ldots, \mathfrak{h}_i, \ldots, \mathfrak{h}_n\}$ be a finite set of hypotheses.
- e a particular datum in the set of evidence, E.
- $\boldsymbol{d}$: $E \to \mathfrak{H}$ maps each $e \in E$ into an element of $\mathfrak{H}$.
- $\boldsymbol{d}^{-1}(\mathfrak{h}_i)$ is the $E_{\mathfrak{h}_i} \subseteq E$ that $\boldsymbol{d}$ maps into $\mathfrak{h}_i$.
- $\ulcorner \boldsymbol{d}(e) \urcorner$ is the name of the $\mathfrak{h} \in \mathfrak{H}$ that $\boldsymbol{d}$ maps e into.[1]

Notation for Normal Distributions

$\mathfrak{N}(\mu, \sigma)$ is a normal distribution with mean $= \mu$ and standard deviation $= \sigma$.
$\mathfrak{N}(\mu, ?)$ is the set of all normal distributions with mean $= \mu$.
$\mathfrak{N}(?, \sigma)$ is the set of all normal distributions with standard deviation $= \sigma$.
$\mathfrak{N}(?, ?)$ is the set of all normal distributions.

[1] Apologies to Quine's ghost for hijacking his notation. In this context, one wants to specify $\mathfrak{h} \in \mathfrak{H}$ before one knows which one it is.

DEFINITION 6.1. Let an *experiment* be a set of persons, perhaps infinite, $\widehat{\mathfrak{H}} = \{\widehat{\mathfrak{h}}_1, \ldots, \widehat{\mathfrak{h}}_i, \ldots\}$ and background information, $\mathfrak{B}$, held to be true by all $\widehat{\mathfrak{h}} \in \widehat{\mathfrak{H}}$. $\mathfrak{B}$ characterizes the experiment and allows the determination of which $e \in E$ obtains.

$\mathfrak{B}$ does not attempt to *define* a "collective" or "chance set-up", but is simply agreed to by all $\widehat{\mathfrak{h}} \in \widehat{\mathfrak{H}}$.

Furthermore, let all persons, $\widehat{\mathfrak{h}} \in \widehat{\mathfrak{H}}$ believe that one of $\mathfrak{h}_{(1,\ldots,n)} \in \mathfrak{H}$ is true and that $\mathfrak{H}$ and $\mathfrak{B}$ determine a set of possible outcomes, $e \in E$. Each e has non-zero probability for some $\mathfrak{h}_i \in \mathfrak{H}$ and In addition, all $\widehat{\mathfrak{h}} \in \widehat{\mathfrak{H}}$ agree on the numeric values of $\mathrm{P}_{\mathfrak{h}_{(1,\ldots,n)}}(e \in E \mid E)$.

Given $\mathfrak{B}$, all of $\widehat{\mathfrak{H}}$ can agree which outcome, $e \in E$, obtains.[2]

POSTULATE I. Any true statement concerning the relationship of a set of hypotheses and evidence must be deductively derived.

AXIOM I.1. Let $e \in E$ be the outcome of an experiment with possible outcomes, E. If one of $\mathfrak{h}_{(1,\ldots,n)} \in \mathfrak{H}$ is true and all but one is contradicted by $e \in E$, that hypothesis is true.

AXIOM I.2. Let $e \in E$ be the outcome of an experiment with possible outcomes, E. If any of $\mathfrak{h}_{(1,\ldots,n)} \in \mathfrak{H}$ might be true and one is true, the truth of any statement about the probability of the truth of $\mathfrak{h}_i$ given $e \in E$, must be deductively derived from the meanings of $\mathfrak{h}_{(1,\ldots,n)} \in \mathfrak{H}$.

AXIOM I.3. Let $e \in E$ be the outcome of an experiment with possible outcomes, E. If any of $\mathfrak{h}_{(1,\ldots,n)} \in \mathfrak{H}$ might be true and one is true, the truth of any statement about the probability the truth of $\mathfrak{h}_i$ given $e \in E$, must be deductively derived from the probability of $e \in E$ given $\mathfrak{h}_{(1,\ldots,n)} \in \mathfrak{H}$, *i.e.* $\mathrm{P}_{\mathfrak{h}_{(1,\ldots,n)}}(e \in E \mid E)$.

DEFINITION 6.2. Let a *decision rule* be a function mapping all $e \in E$ onto some $\mathfrak{h} \in \mathfrak{H}$, $\boldsymbol{d}\colon E \to \mathfrak{H}$ and let $E_{\mathfrak{h}_i}$ be the subset of E mapped onto $\mathfrak{h}_i$. The occurrence of a member of $E_{\mathfrak{h}_i}$ causes $\boldsymbol{d}$ to choose $\mathfrak{h}_i$

POSTULATE II. In choosing a decision rule, one ought to choose the rule with the highest probability of being correct.

DEFINITION 6.3. Let the *acuity*, $\boldsymbol{\mathcal{A}}_{\boldsymbol{d}}$, of a decision rule, $\boldsymbol{d}\colon E \to \mathfrak{H}$, be

$$\boldsymbol{\mathcal{A}}_{\boldsymbol{d}} = \min_{\mathfrak{h}_i=1}^{n}\Big(\mathrm{P}_{\mathfrak{h}_i}(E_{\mathfrak{h}_i} \mid E)\Big)$$

The acuity is based on the probability $\mathfrak{h}_i$ gives the evidence, $E_{\mathfrak{h}_i}$, that $\boldsymbol{d}$ maps into $\mathfrak{h}_i$.

Now if all parties, $\widehat{\mathfrak{H}}$, agree to use a decision rule, $\boldsymbol{d}\colon E \to \mathfrak{H}$ with acuity $= \boldsymbol{\mathcal{A}}_{\boldsymbol{d}}$, then all must agree to this statement:

> "If one of $\mathfrak{H}$ is true, the probability that $\ulcorner \boldsymbol{d}(\mathrm{e}) \urcorner$ is true is at least $\boldsymbol{\mathcal{A}}_{\boldsymbol{d}}$ and the probability of some other member of $\mathfrak{H}$ being true is at most $1 - \boldsymbol{\mathcal{A}}_{\boldsymbol{d}}$."

where e is the outcome of the experiment and each $\widehat{\mathfrak{h}} \in \widehat{\mathfrak{H}}$ uses *probability* to mean what every he wishes, so long as it follows the axioms of probability.

In many cases, the statement will be:

> "If one of $\mathfrak{H}$ is true, the probability that $\ulcorner \boldsymbol{d}(\mathrm{e}) \urcorner$ is true is *exactly* $\boldsymbol{\mathcal{A}}_{\boldsymbol{d}}$ and the probability of some other member of $\mathfrak{H}$ being true is *exactly* $1 - \boldsymbol{\mathcal{A}}_{\boldsymbol{d}}$."

[2]It is not uncommon to see: $\mathrm{P}(e \mid \mathfrak{h}_i)$ or, more properly, $\mathrm{P}(e \mid \mathfrak{h}_i\,\mathfrak{B})$. The usage is incorrect. It is not sufficient for something to be true for it to be placed in the conditioning position in a conditional probability. Almost all of the puzzles in the interpretation of probability come from confusing what would be appropriate in one probability system with what is inappropriate in another. In this case, all $e \in E$ are outcomes; $\mathfrak{h}_i$ and $\mathfrak{B}$ are not. One must be very precise in the meanings of what goes inside $\mathrm{P}(\cdot \mid \cdot)$.

This is an exceptionally powerful result! $\widehat{\mathfrak{H}}$ are forced to unanimity even if their hypotheses are incommensurable in areas other than, that $\mathfrak{B}$ allows the recognition of e. One person may be defining probability as a propensity, another, relative frequency, and another, degree of belief.

The procedure can be expanded to a countably infinite number of hypotheses and give meaningful results if the hypotheses are "far enough" from each other.[3]

If the hypotheses have the cardinality of the real numbers, they will always be too close together and if treated as independent hypotheses, $\boldsymbol{\mathcal{A}_d} = 0$.[4]

The procedure can be extended to sets of hypotheses in the following way.

DEFINITION 6.4. A set of hypotheses, $\mathfrak{H}$, is *incompletely specified* relative to $\mathfrak{B}$ if there are one or more parameters unspecified such that:

$$\text{for } \mathfrak{h} \in \mathfrak{H} \text{ and } e \in E,\ \mathrm{P}_{\mathfrak{h}}(e \mid E) \text{ cannot be calculated.}$$

An example would be $\mathfrak{N}(\mu, ?)$.

DEFINITION 6.5. *Completing* a set of hypotheses, $\mathfrak{H}$, incompletely specified relative to $\mathfrak{B}$, is the selection of $\mathfrak{h} \in \mathfrak{H}$ so that:

$$\mathrm{P}_{\mathfrak{h}}(e \mid E) \text{ can be calculated for all } e \in E.$$

POSTULATE III. When *completing* a set of hypotheses, $\mathfrak{H}$, incomplete relative to $\mathfrak{B}$, given evidence, $e \in E$, one should choose $\mathfrak{h} \in \mathfrak{H}$ such that $\mathrm{P}_{\mathfrak{h}}(e \mid E)$ is maximized.[5]

DEFINITION 6.6. Let $\mathfrak{H} = \{\mathfrak{H}_1, \ldots, \mathfrak{H}_i, \ldots, \mathfrak{H}_n\}$ be a set of sets of hypotheses, then let the *hypothesis selection function* of $\boldsymbol{s}\colon E \to \mathfrak{H}_{(1,\ldots,n)} \in \mathfrak{H}$ such that:

$$\boldsymbol{s}(e \in E) = \left\{\mathfrak{h}'_1 \in \mathfrak{H}_1, \ldots, \mathfrak{h}'_i \in \mathfrak{H}_i, \ldots, \mathfrak{h}'_n \in \mathfrak{H}_n\right\} = \mathfrak{H}'$$

where, for $1 \le i \le n$,

$$\mathfrak{h}'_i \in \mathfrak{H}_i \text{ such that } \mathrm{P}_{\mathfrak{h}'_i}(e \mid E) = \max_{\mathfrak{h} \in \mathfrak{H}_i} \mathrm{P}_{\mathfrak{h}}(e \mid E)$$

Let the *acuity*, $\boldsymbol{\mathcal{A}_d}$ of a decision rule, $\boldsymbol{d}\colon E \to \mathfrak{H}$ be the *acuity* of $\boldsymbol{d'}\colon E \to \mathfrak{H}'$

$$\boldsymbol{\mathcal{A}_{d'}} = \min_{\mathfrak{h}'_{i=1}}^{n}\left(\mathrm{P}_{\mathfrak{h}'_i}(E_{\mathfrak{h}'_i} \mid E)\right)$$

In other words, from each set of hypotheses, choose the one that gives the highest likelihood of obtaining the evidence, e, that actually occurs and use the set of representative hypotheses to calculate the acuity of a decision rule from $E \to \{\mathfrak{h}'_{(1,\ldots,n)}\}$.

DEFINITION 6.7. A Maximum Acuity Decision Rule, $\boldsymbol{d}\colon E \to \mathfrak{H}$, is the mapping that maximizes acuity, $\boldsymbol{\mathcal{A}_d}$.

---

[3]Suppose there is a bean whose weight is follow a normal distribution with mean $=1$ and standard deviation $=1$, $\mathfrak{N}(1,1)$. Further suppose that a bag contains 1 or 2 or 3 or ... beans and one wants to determine the number of beans by weighing the bag. The hypotheses are $\{\mathfrak{N}_1(1,\sqrt{1}), \mathfrak{N}_2(2,\sqrt{2}), \mathfrak{N}_3(3,\sqrt{3}), \ldots \mathfrak{N}_n(n,\sqrt{n}), \ldots\}$. In this case, the distance between the means stays constant while the standard deviation approaches infinity as n approaches infinity. In this case, $\boldsymbol{\mathcal{A}_d} \to 0$.

Suppose the bags could contain only a number of beans in the Fibonacci series, leaving off the first two numbers in the series. Then the hypotheses would be $\{\mathfrak{N}_1(1,\sqrt{1}), \mathfrak{N}_2(2,\sqrt{2}), \mathfrak{N}_3(3,\sqrt{3}), \mathfrak{N}_5(5,\sqrt{5}), \ldots \mathfrak{N}_n(n_{-2}+n_{-1}, \sqrt{n_{-2}+n_{-1}}), \ldots\}$. In this case, the acuity would be mostly determined by the first two hypotheses; $\boldsymbol{\mathcal{A}_d} \approx .65$.

[4]This lends support to Popper's contention that, in the Bayesian context, because they are so numerous, the only proper "prior probability" for a hypothesis is zero.

[5]This seems innocuous. It is simply letting $\mathfrak{H}$ put its best foot foreward. However, when the implications of the postulate become clear, the waters of the soft sciences will be roiled.

DEFINITION 6.8. An Equal Acuity Decision Rule, $\boldsymbol{d}\colon E \to \mathfrak{H}$, is the mapping such that $\forall \mathfrak{h}_i \in \mathfrak{H}$, $\mathrm{P}_{\mathfrak{h}_i}(\boldsymbol{d}^{-1}(\mathfrak{h}_i) \mid E) = k$. That is, the probability of the evidence that would cause $\boldsymbol{d}$ to choose $\mathfrak{h}_i$ is equal to k for each $\mathfrak{h}_i$.

THEOREM 6.1. If one exists, an Equal Acuity Decision Rule is a Maximum Acuity Decision Rule.

Proof:

Since acuity is a function of minimum of the probability of the evidence that is mapped onto each of the hypotheses and all evidence is mapped onto some hypothesis, any change that increases the probability of choosing one hypothesis must reduce the probability of some other—thus reducing the minimum. *Q. E. D.*

In the all the ordinary cases, there *is* an optimum decision rule. By agreeing to use the optimum decision rule, all $\widehat{\mathfrak{H}}$ will forced into agreement by the weight of the evidence. The statement that must be made after the decision rule is applied to the evidence is that if any of the hypotheses is true, no matter which one, there is a probability, equal to the acuity, of the choice being correct.

It is important to see clearly what does *not* affect the acuity of a decision rule. What $\widehat{\mathfrak{H}} = \{\widehat{\mathfrak{h}}_1, \ldots, \widehat{\mathfrak{h}}_i, \ldots\}$ might feel about $\mathfrak{h}_{(1,\ldots,n)} \in \mathfrak{H}$ does not matter. That is subsumed by "based on the occurence of e." How $\mathfrak{h}_{(1,\ldots,n)} \in \mathfrak{H}$ were chosen and the timing of their choice does *not* matter. They can be the result of data mining and they can be included after the occuence of e. Both situations are subsumed by "If one of $\mathfrak{h}_{(1,\ldots,n)} \in \mathfrak{H}$ is true."

We will now work out the details for most of the standard situations met with in statistical testing. In general, there is no analytic method for deriving the maximum acuity decision rule. For continuous probability distributions, the general principle is that the integral of the probability density mapped into each hypothesis is equal for each hypothesis and maximal. Many situations can be reduced to the choice between two normal distributions.

On the basis of an experimental sample, hypothesized to be from a normal distribution, we will construct a Maximum Acuity Decision Rule to choose between:

1. Two hypothesized normal distributions with specific means and standard deviations
   - The standard deviations can be equal and the means unequal
   - The means can be equal and the standard deviations unequal
   - Both the means and standard deviations can be unequal
2. Two hypothesized normal distributions, one with specific mean and standard deviation, the other with specified mean and unspecified standard deviation
3. Two hypothesized normal distributions with specific means and unspecified standard deviations
4. Two hypothesized normal distributions, one with specified mean and unspecified standard deviation, the other with unspecified mean and unspecified standard deviation

In the case of two normal distributions (and one assumes, two unimodal differentiable distributions), the maximum acuity decision rule is defined by two cuts, a and b, that surround the "pointier" of the distributions so that the integral of the probability density of that distribution between a and b is equal to the integral of the probability density of the other distribution outside (a, b). In the special case where the standard deviations are equal, the decision rule reduces to one cut midway between the means.

In the following examples, Mathematica's arbitrary precision arithmetic was used to find a and b such that the abovementioned integrals are equal and that the difference in probability density between the two distributions are equal at a and b. Thus moving a and b toward each other, moving both away from each other, moving both to the right, or moving both to the left steals probability from one distribution and gives it to the other, reducing acuity, which is driven by the minimum.

Conclusion:

- If one of a set of hypotheses, $\mathfrak{h}_{(1,\ldots,i,\ldots,n)} \in \mathfrak{H}$, is true and
- given agreement on background conditions, $\mathfrak{B}$, that allows the determination of which evidence, $e \in E$ is the case, and
- if there exists an equal acuity decision rule, $\boldsymbol{d}$, that maps e into $\mathfrak{H}$,
- then if $\boldsymbol{d}$: $e \to \mathfrak{h}_j$,
- the probability that $\mathfrak{h}_j$ is true equals $\boldsymbol{\mathcal{A}_d}$ and the probability that $\mathfrak{h}_{i \neq j}$ is true is equal to $1 - \boldsymbol{\mathcal{A}_d}$.

CHAPTER 7

Two Normal Distributions with Equal Standard Deviations

Notation of σ and s

When a normal distribution with mean $= \mu$ or a sample with $= \mathbf{m}$ is discussed, σ means σ_μ and s means s_m.

Statisticians have been looking at diagrams like those that we use to illustrate Sindbad's method since they first were introduced to statistics. Traditionally, the diagrams have been used to talk about the "null" hypothesis—to avoid accepting when it is false or rejecting it when it is true. Our purpose is different. Our diagrams illustrate a rule that leads to probable truth.

This is just like the puzzles that were presented to Sindbad, but rather than observing what coin or gem that was drawn, the task is to observe the location of the mean of a sample. One of two normal distributions is assumed to be true. They have unequal means and equal standard deviations. The rule consists of a cut midway between the two means. The decision rest on which side of the cut the sample mean falls. If the sample mean falls exactly on the cut, there is a rule, but no decision.

	Unequal μ, Equal σ The Decision Rule, d, Is Determined by One Cut at a	
	$m < a$	$m > a$
$\mathfrak{h}_1$: Whose $\mu < a$	$\mathcal{A}_d$	$1 - \mathcal{A}_d$
$\mathfrak{h}_2$: Whose $\mu > a$	$1 - \mathcal{A}_d$	$\mathcal{A}_d$
	say: **"$\mathfrak{h}_1$ is true"**	**say:** **"$\mathfrak{h}_2$ is true"**
	The Decision: Acuity $= \mathcal{A}_d$	

TABLE 7.1. Unequal μ, Equal σ

In Figure 7.1 below, there is a probability $= .691462$ of being correct, if one chooses $\mathfrak{N}(1,1)$ when the sample mean is greater than .5 and $\mathfrak{N}(0,1)$ when it is less than .5.

From this example, it may seem that one is asserting "$\mathfrak{N}(1,1)$" or "$\mathfrak{N}(0,1)$" based on which distribution has the higher likelihood of the sample mean occurring. This is not the case.

The choice is made by partitioning the possible values of the sample mean into two sets such that the integral of the probability density is maximized and equal for both distributions. If $\mathfrak{N}(1,1)$ is true, there is a probability $= .691462$ that the mean will be to the right of .5; If $\mathfrak{N}(0,1)$ is true, there is a probability $= .691462$ that the mean will be to the left of 0.5. One can unequivocally make the statement: "If either $\mathfrak{N}(1,1)$ or $\mathfrak{N}(0,1)$ is true and I choose based on whether the sample mean fall the right or left of 0.5, I have a .691462 probability of being correct."

As μ of the first normal distribution progresses from 1 to 2, to 3, to 4, to 5, acuity progresses from 0.691462 to 0.841345 to 0.933193 to 0.97725 to 0.99379.

This is the usual situation in the physical sciences. Rival theories have a numerical predictions for the outcome of a particular experiment and variation in results is the accounted for by the accuracy of the instrumentation, thus equally applicable to the predictions of each of the theories. An example can be found in Chapter 19 dealing with General Relativity, Special Relativity, and Newtonian optics.

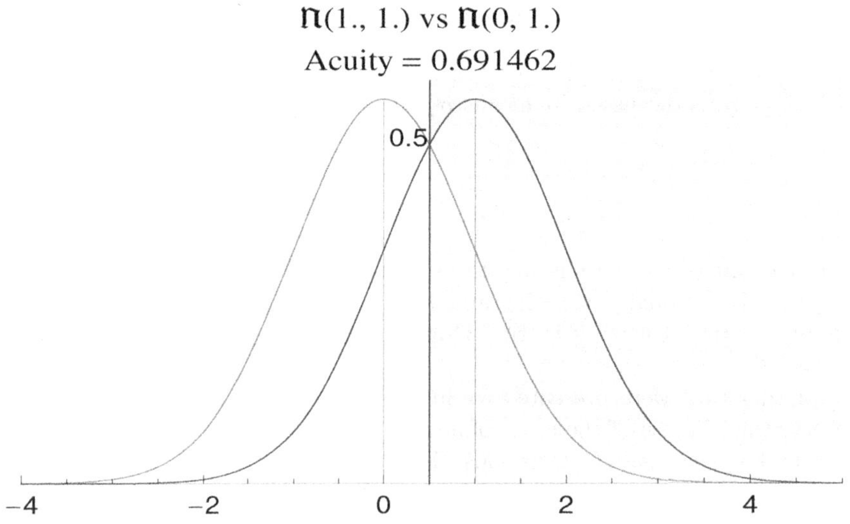

FIGURE 7.1. Maximum Acuity Decision Rule for Two Normal Distributions, $\mathfrak{N}(0,1)$ and $\mathfrak{N}(1,1)$

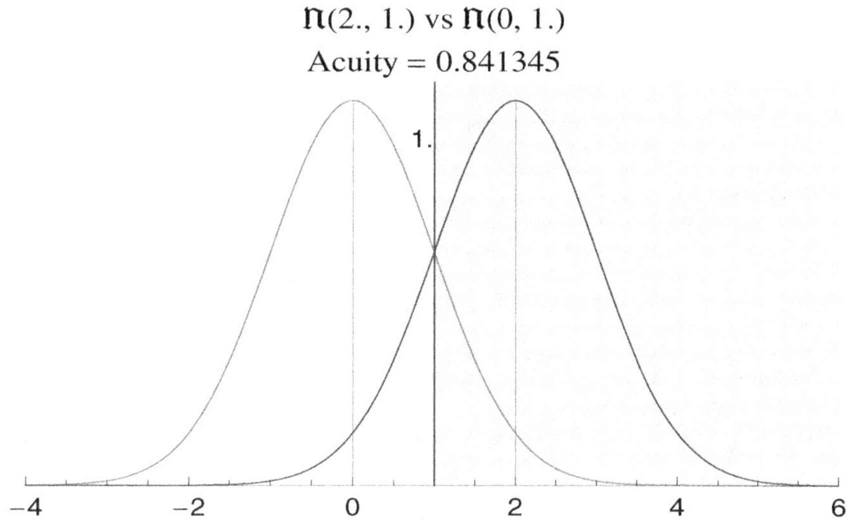

FIGURE 7.2. Maximum Acuity Decision Rule for Two Normal Distributions, $\mathfrak{N}(0,1)$ and $\mathfrak{N}(2,1)$

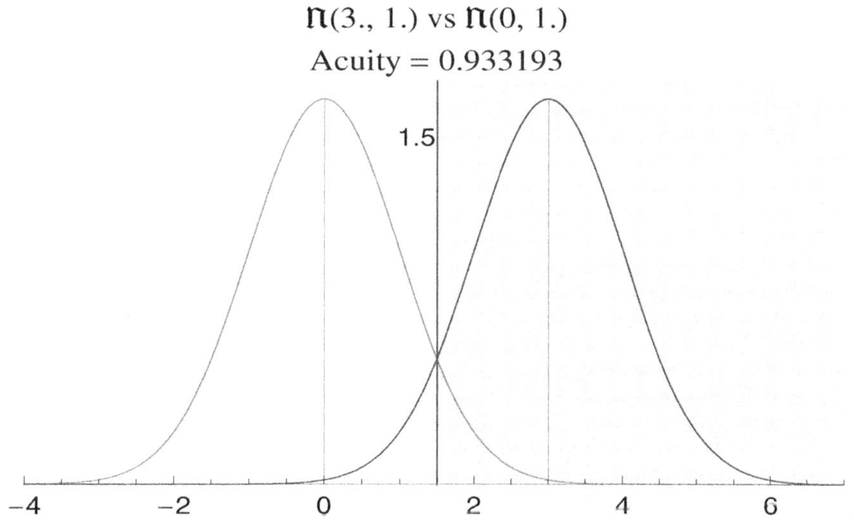

FIGURE 7.3. Maximum Acuity Decision Rule for Two Normal Distributions, $\mathfrak{N}(0,1)$ and $\mathfrak{N}(3,1)$

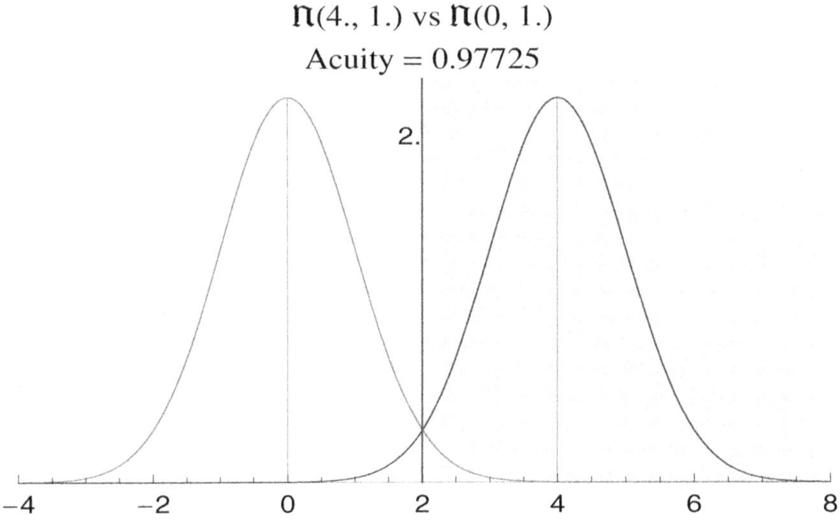

FIGURE 7.4. Maximum Acuity Decision Rule for Two Normal Distributions, $\mathfrak{N}(0,1)$ and $\mathfrak{N}(4,1)$

Figure 7.5 shows the situation that is considered the gold standard in the physical sciences. The equal standard deviations are produced by the instrumentation. If one assumes that the leftmost normal distribution is the "null" hypothesis and the rightmost normal distribution is the result of an experiment, so that $m = \mu_2$ and $s = \sigma_1$, the hard sciences would consider the leftmost distribution falsified. Of course, all this is a misstatement. The "null" distribution in the hard sciences would be a very specific hypothesis and, try as it might, an ordinary probability statement cannot falsify an hypothesis in the logical sense. What one *can* say, is that if one or the other of these hypothesis is true and the sample mean is to the left of the cut, there is less than a 1 to 160 chance that the first hypothesis is true and, if the right of the cut, a 160 to 1 chance that the second is true.

Practitioners in the soft sciences imagine that this is what *they* are doing. This is not the case. Chapters 12 and 13 show what *is* the case.

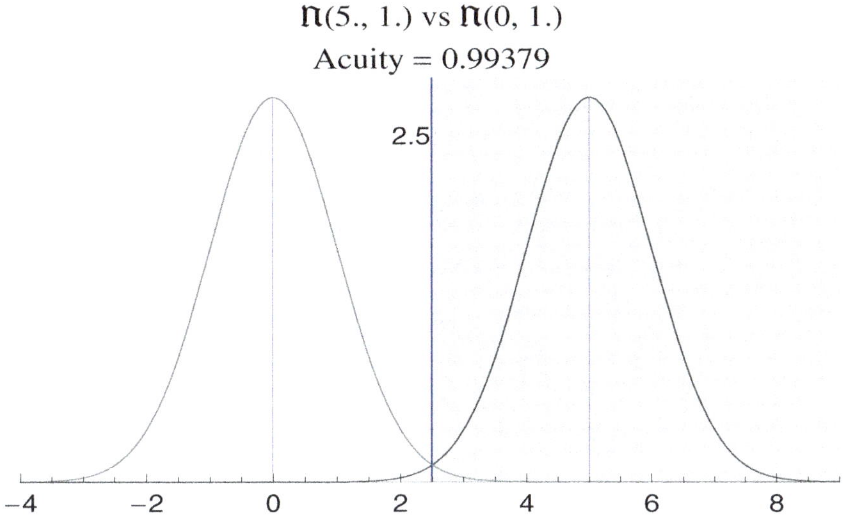

FIGURE 7.5. Maximum Acuity Decision Rule for Two Normal Distributions, $\mathfrak{N}(0,1)$ and $\mathfrak{N}(5,1)$

Figure 7.6 shows how the separation of the means, scaled in terms of the common standard deviation, is reflected in the acuity of the decision rule. Acuity rises from .5 at separation 0 (the degenerate case) to 0.99379 at separation 5. On the right side of the graph, at separation of 15, the odds of deciding correctly are 3.1275×10^{13} to one. This rapid ascent because there is separation of the means based on real theory and no interplay between the separation of the theoretical means and standard deviations. Cases where there is an interplay will be dealt with beginning in Chapter 10.

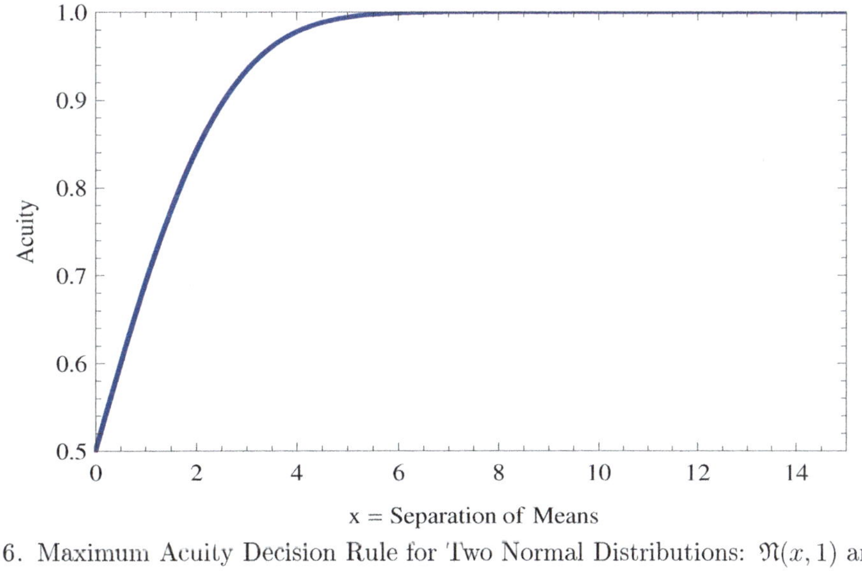

FIGURE 7.6. Maximum Acuity Decision Rule for Two Normal Distributions: $\mathfrak{N}(x,1)$ and $\mathfrak{N}(0,1)$

CHAPTER 8

Two Normal Distributions with $\mu_1 = \mu_2$ & $\sigma_1 \neq \sigma_2$

It may be hard to associate this type of decision with a situation in scientific research and to disassociate it from a situation that sounds similar, but is not the same.

First the disassociation. This decision has nothing to do with the question of whether two normal distributions have equal standard variations.

Second, in the hard sciences, one would never find that the choice of hypothesis would effect the the error of the instrumentation. In the soft sciences, standard deviations are never hypothesized *a priori.* For completeness sake, in this chapter and the next, we analyze the situation where both means and standard deviations are *given* and the standard deviations are unequal.

The question *is*, "If one of these two distributions is the true distribution of the mean of the sample I am about to draw, what is the decision rule that gives me the highest probability of being correct?"

	Equal μ, Unequal σ The Decision Rule, $\boldsymbol{d}$, Is Determined by Two Cuts at $\boldsymbol{a}$ & $\boldsymbol{b}$	
	$a < m > b$	$m < a$ or $m > b$
$\mathfrak{h}_1$: With the smaller σ	$\mathcal{A}_d$	$1 - \mathcal{A}_d$
$\mathfrak{h}_2$: With the larger σ	$1 - \mathcal{A}_d$	$\mathcal{A}_d$
	say: "$\mathfrak{h}_1$ is true"	say: "$\mathfrak{h}_2$ is true"
	The Decision: Acuity $= \mathcal{A}_d$	

TABLE 8.1. Equal μ, Unequal σ

Likelihood

In the previous chapter, the Maximum Acuity Decision Rule always chose the hypothesis with the highest likelihood having produced the sample. Here, in order to maximize the probability of choosing correctly, one must sometimes choose the hypothesis less likely to have produced the sample. This drives a spike through the heart of the THE LIKELIHOOD AXIOM in Edwards [**10**, Chapter 3.3, page 32].

If the sample mean falls outside the area between the two cuts, choose the hypothesis with the larger standard deviation; otherwise choose that with the smaller standard deviation. Note the little "wings" to the outside of the cuts in Figure 8.1. These are areas where one must choose the less likely hypothesis in order to maximize acuity.

The decision rules can be characterized by the ratio of the two standard deviations. When the ratio progresses from 2 to 3, 4, and 5; acuity goes from .644258 to .71921, .766042, and .798409. As the ratio increases the decision cuts move equally farther from the common mean.

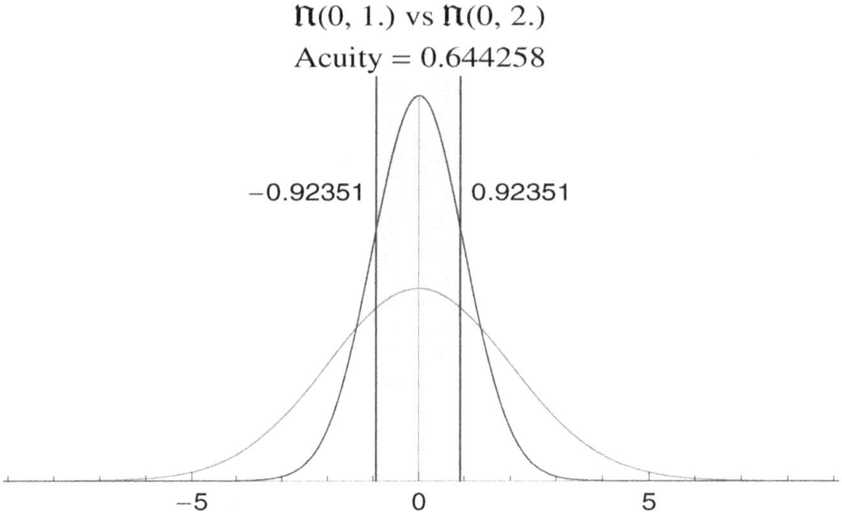

FIGURE 8.1. Maximum Acuity Decision Rule for Two Normal Distributions, $\mathfrak{N}(0,1)$ and $\mathfrak{N}(0,2)$

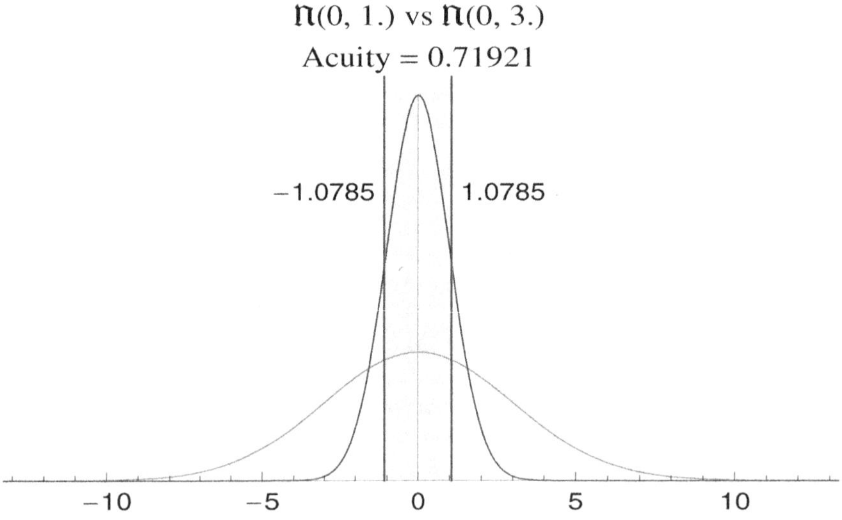

FIGURE 8.2. Maximum Acuity Decision Rule for Two Normal Distributions, $\mathfrak{N}(0,1)$ and $\mathfrak{N}(0,3)$

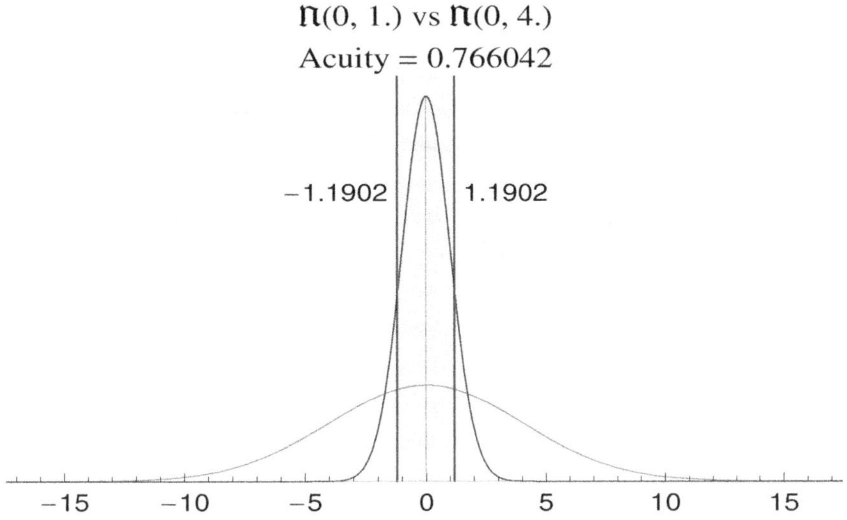

FIGURE 8.3. Maximum Acuity Decision Rule for Two Normal Distributions, $\mathfrak{N}(0,1)$ and $\mathfrak{N}(0,4)$

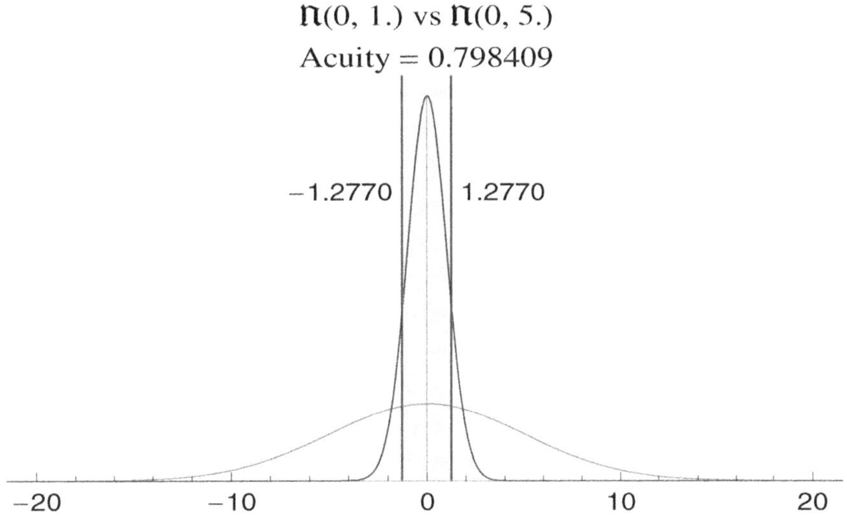

FIGURE 8.4. Maximum Acuity Decision Rule for Two Normal Distributions, $\mathfrak{N}(0,1)$ and $\mathfrak{N}(0,5)$

Figure 8.5 plots acuity vs. the standard deviation ratio. The acuity rises fairly quickly from .5, but then levels off at fairly low level so that huge differences in the standard deviation ratio make almost no difference in acuity. This is because all of the action takes place in the tails.

In an extreme example, at the very right of the graph, deciding between $\mathfrak{N}(0,1)$ and $\mathfrak{N}(0,15)$, the odds of being right are only 10.112 to 1, the acuity only 0.910007.

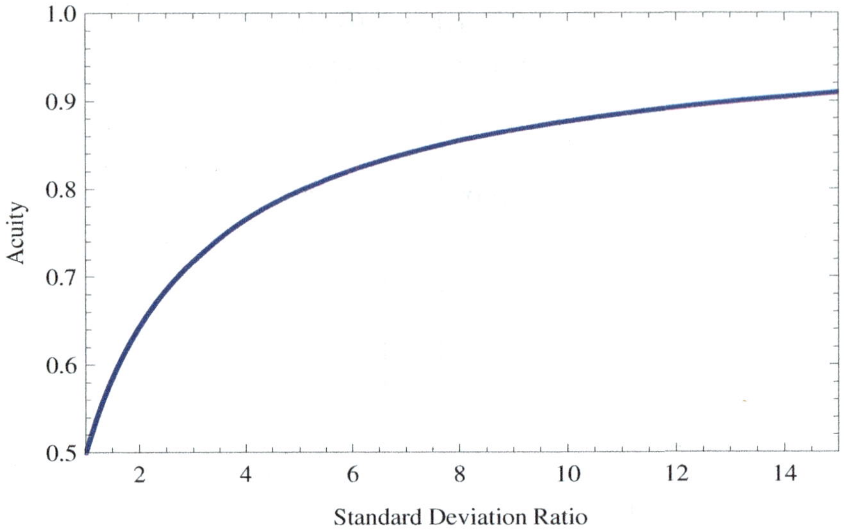

FIGURE 8.5. Maximum Acuity for Two Normal Distributions with equal μ where the x-axis $= \frac{\sigma_1}{\sigma_0}$

CHAPTER 9

Two Normal Distributions with $\mu_1 \neq \mu_2$ & $\sigma_1 \neq \sigma_2$

It can occur in various sciences that there are two hypotheses that differ in mean and standard deviation. It is not usually the case in the hard sciences, but it can occur. A process could be hypothesized to be controlled by two disparate processes that have different outcomes and different dispersions of their results. In that case, the experimental sample is deciding between two well-formed hypotheses.

In the soft sciences, the source of the two hypotheses *could* be the result of the previous experiments. But in the soft sciences, one seldom finds one well-formed hypothesis, let alone two.

Figure 9.1 is similar to an old friend from Figure 7.5. The standard deviations are equal in the cut is midway between the means.

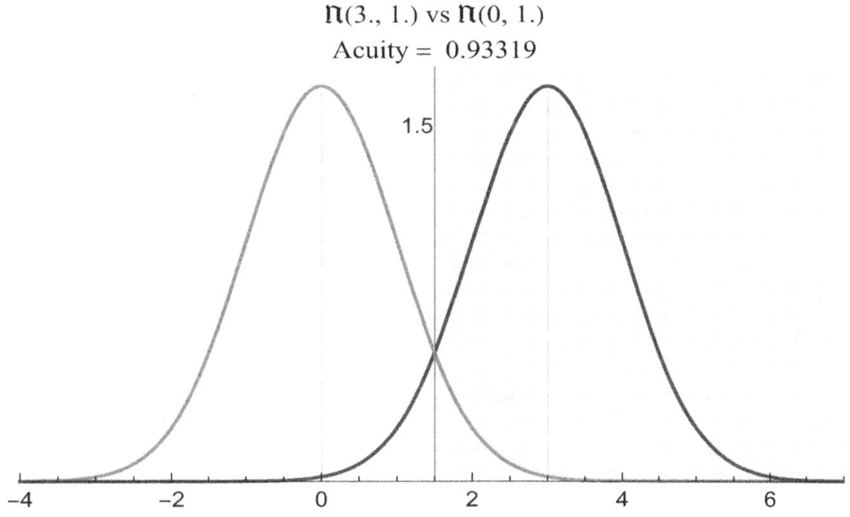

FIGURE 9.1. Maximum Acuity Decision Rule for Two Normal Distributions, $\mathfrak{N}(3, 1)$ and $\mathfrak{N}(0, 1)$

The decision procedure when the σs are not equal is shown in the table below. When the sample mean falls between the cuts the decision goes to the distibution whose μ is between the cuts, otherwise, it goes to the distribution whose μ is outside the cuts.

	Unequal μ & Unequal σ The Decision Rule, d, Is Determined by Two Cuts at a & b	
	$a < m > b$	$m < a$ or $m > b$
$\mathfrak{h}_1$: Whose μ is between the cuts	$\mathcal{A}_d$	$1 - \mathcal{A}_d$
$\mathfrak{h}_2$: Whose μ is outside the cuts	$1 - \mathcal{A}_d$	$\mathcal{A}_d$
	say: "$\mathfrak{h}_1$ is true"	say: "$\mathfrak{h}_2$ is true"
	The Decision: Acuity $= \mathcal{A}_d$	

TABLE 9.1. Unequal μ & Unequal σ

In the following charts, acuity acts in a way unlike the previous examples. As σ_2 increases, $1, 2, 2.5, 3$, acuity declines, $0.93319, 0.83213, 0.80800, 0.80056$. Then there is a change. As σ_2 increases, $3.5, 4, 5$, acuity rises, $0.80199, 0.80753, 0.82247$.

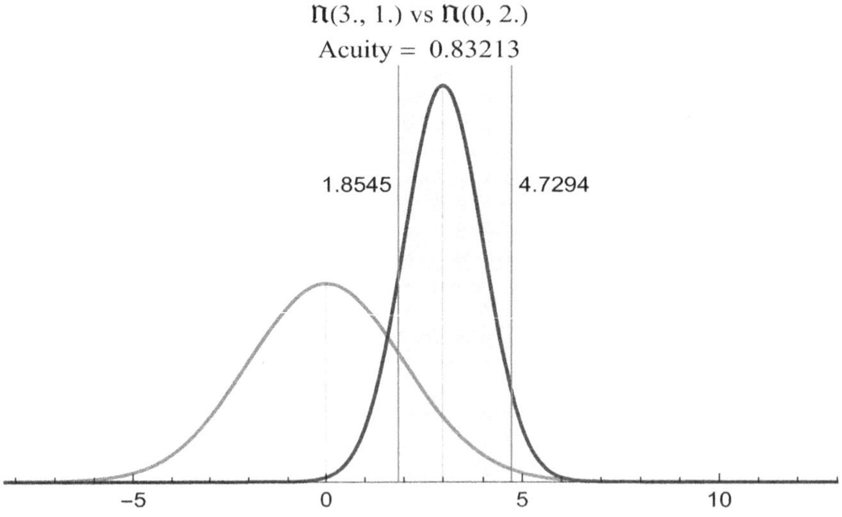

FIGURE 9.2. Maximum Acuity Decision Rule for Two Normal Distributions, $\mathfrak{N}(3,1)$ and $\mathfrak{N}(0,2)$

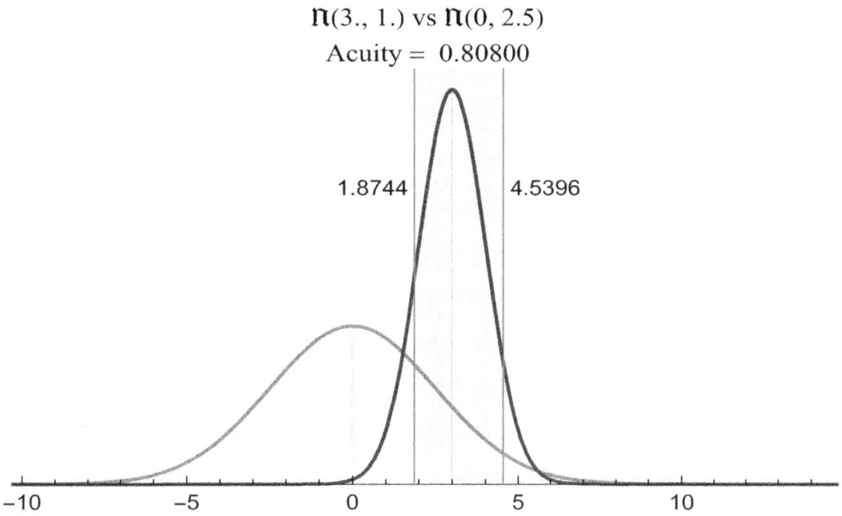

FIGURE 9.3. Maximum Acuity Decision Rule for Two Normal Distributions, $\mathfrak{N}(3,1)$ and $\mathfrak{N}(0,2.5)$

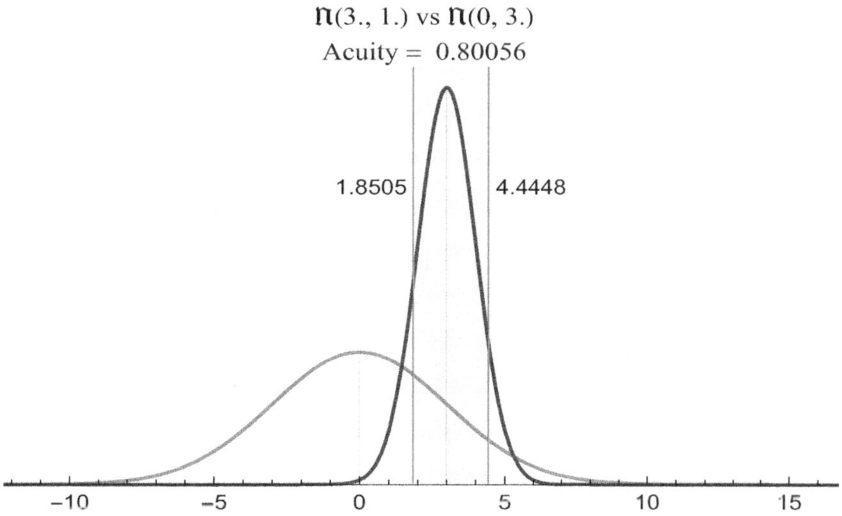

FIGURE 9.4. Maximum Acuity Decision Rule for Two Normal Distributions, $\mathfrak{N}(3,1)$ and $\mathfrak{N}(0,3)$

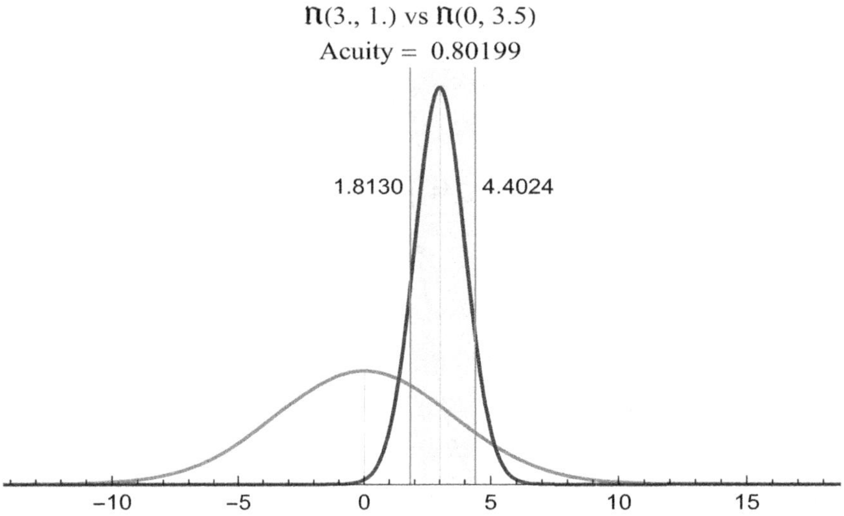

FIGURE 9.5. Maximum Acuity Decision Rule for Two Normal Distributions, $\mathfrak{N}(3,1)$ and $\mathfrak{N}(0,3.5)$

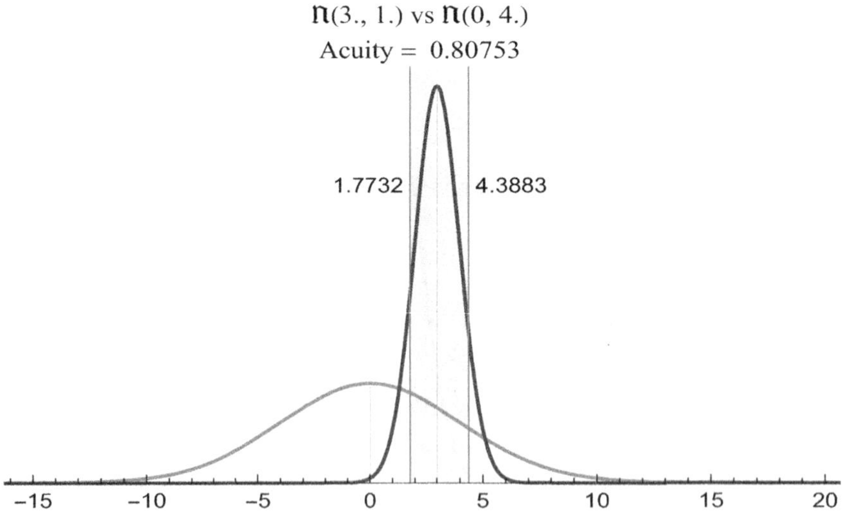

FIGURE 9.6. Maximum Acuity Decision Rule for Two Normal Distributions, $\mathfrak{N}(3,1)$ and $\mathfrak{N}(0,4)$

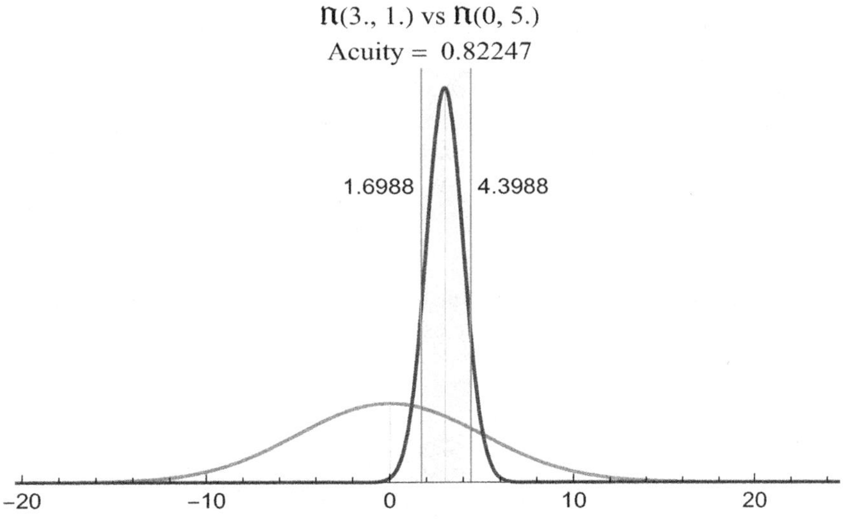

FIGURE 9.7. Maximum Acuity Decision Rule for Two Normal Distributions, $\mathfrak{N}(3,1)$ and $\mathfrak{N}(0,5)$

One can make sense of this behavior by way of a three dimensional graph.

Figure 9.8 shows the interaction of the separation of the means and the ratio of the standard deviations. Everything is scaled so that the smaller standard deviation = 1.

The SW edge of the surface is the acuity when $\mu_1 = \mu_2$. It is identical to Figure 8.5. The SE edge of the surface is the acuity when $\sigma_1 = \sigma_2$. It is identical to Figure 7.6.

When one travels from the SW edge to the NE edge, parallel to the SE edge, acuity is monotone increasing. When one travels from the SE edge to the NW edge, parallel to the SW edge, acuity dips to a low when

$$\frac{|\mu_1 - \mu_2|}{\min(\sigma_1, \sigma_2)} = \frac{\max(\sigma_1, \sigma_2)}{\min(\sigma_1, \sigma_2)}$$

and then is monotone increasing. The previous seven charts are points on a line that starts halfway up the SE edge at $\mu_1 = 3$ and $\sigma_2 = 1$ and heads NW parallel to the SW edge as σ_2 increases to 5.

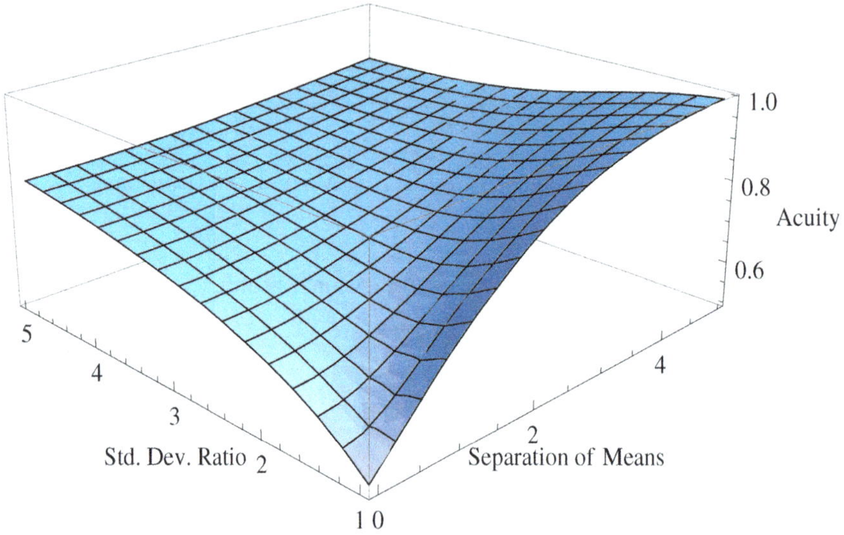

FIGURE 9.8. Acuity of $\mathfrak{N}(0, 1)$ vs $\mathfrak{N}(\mu, \sigma)$

When the ratio of σ_1 to σ_2, increasing the spread between μ_1 and μ_2 uniformly increases the acuity of the optimum decision rule. Except when $\sigma_1 = \sigma_2$, when one travels from the SE to the NW there is a dip in acuity.

Much more interesting things happen when the sample guides which individual member of a family of distributions is one of the two hypotheses put to the test. The next chapter shows the result.

CHAPTER 10

$\mathfrak{N}(\mu_1, \sigma_1)$ or $\mathfrak{N}(\mu_2, ?)$ hypothesized, Sample = $\mathfrak{N}(m, s)$

In the following examples, two normal distributions with μ_1, σ_1, & μ_2 given, but σ_2 unknown. A sample is taken. The evidence is the sample's mean, m. The standard deviation of the mean is s. The problem is to find the value of σ_2 consistent with the sample, $(m,\ s)$ and the assumption that the sample was drawn from a normal distribution with mean $= \mu_2$. The best candidate is:

$$\sigma_2 = \sqrt{s^2 + (\mu_1 - m)^2}$$

The maximum acuity decision rule and resultant acuity is calculated for $\mathfrak{N}(\mu_1, \sigma_1)$ and $\mathfrak{N}(\mu_2, \sigma_2)$ as if they were hypothesized in the beginning.[1]

Figure 10.1 shows the start of a progression as the sample mean marches to the right past the fixed $\mathfrak{N}(\mu_2, \sigma_2)$. The sample mean $m = \mu_1$ so the sample standard deviation, s is the only driver for σ_1 thus $\mu_2 = m$. Since $\sigma_1 = \sigma_2$, the maximum acuity decision rule is a cut at the mid-point between μ_1 and μ_2.

With a difference of 5 times the mutual standard deviation the acuity of the decision rule is quite robust, .9937903. This, again, is the standard in the hard sciences, but it depends on the happenstance that $m = \mu_1$.

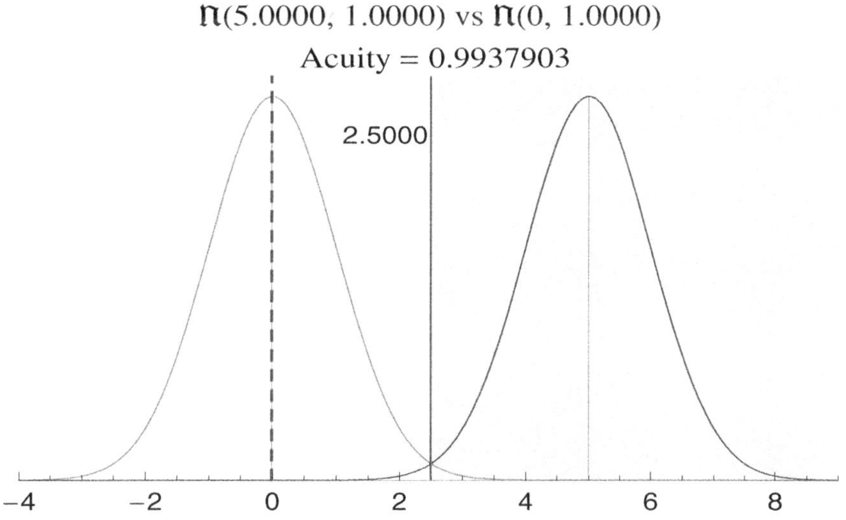

FIGURE 10.1. Maximum Acuity Decision Rule for Two Normal Distributions: $\mathfrak{N}_1(5,1)$ & $\mathfrak{N}_2(0,?)$ & Sample = $\mathfrak{N}(0,1)$

[1]It is helpful to think of every experiment as a replication. The assumption that the rival distributions are hypothesized erases any meaningful difference between an original experiment and a replication.

	μ_1, μ_2 & σ_2 Known, σ_1 Unknown The Decision Rule, d, Is Determined by Two Cuts at a & b	
	$a < m > b$	$m < a$ or $m > b$
$\mathfrak{h}_1$: Whose μ is between the cuts	$\mathcal{A}_d$	$1 - \mathcal{A}_d$
$\mathfrak{h}_2$: Whose μ is outside the cuts	$1 - \mathcal{A}_d$	$\mathcal{A}_d$
	say: "$\mathfrak{h}_1$ is true"	say: "$\mathfrak{h}_2$ is true"
	The Decision: Acuity $= \mathcal{A}_d$	

TABLE 10.1. μ_1, μ_2 & σ_2 Known, σ_1 Unknown

Figure 10.2 starts the march to the right. σ_1 begins to grow larger because of influence of $(m - \mu_1)^2$ causing the acuity to decline. The decision rule produces two cuts surrounding μ_2. m is still outside the band so the decision goes to $\mathfrak{N}(\mu_1, \sigma_1)$. This persists until somewhere between $m = 3$ in Figure 10.4 and $m = 4$ in Figure 10.5.

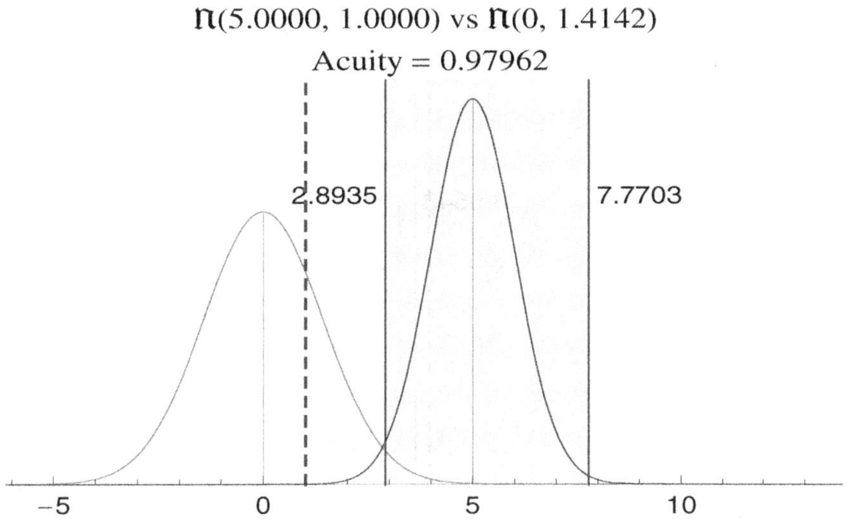

FIGURE 10.2. Maximum Acuity Decision Rule for Two Normal Distributions: $\mathfrak{N}_1(5,1)$ & $\mathfrak{N}_2(0,?)$ & Sample = $\mathfrak{N}(1,1)$

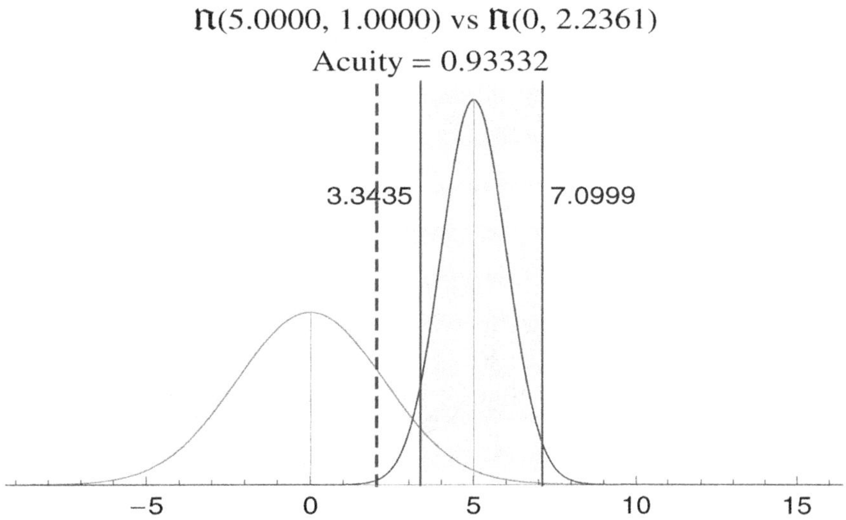

FIGURE 10.3. Maximum Acuity Decision Rule for Two Normal Distributions: $\mathfrak{N}_1(5,1)$ & $\mathfrak{N}_2(0,?)$ & Sample = $\mathfrak{N}(2,1)$

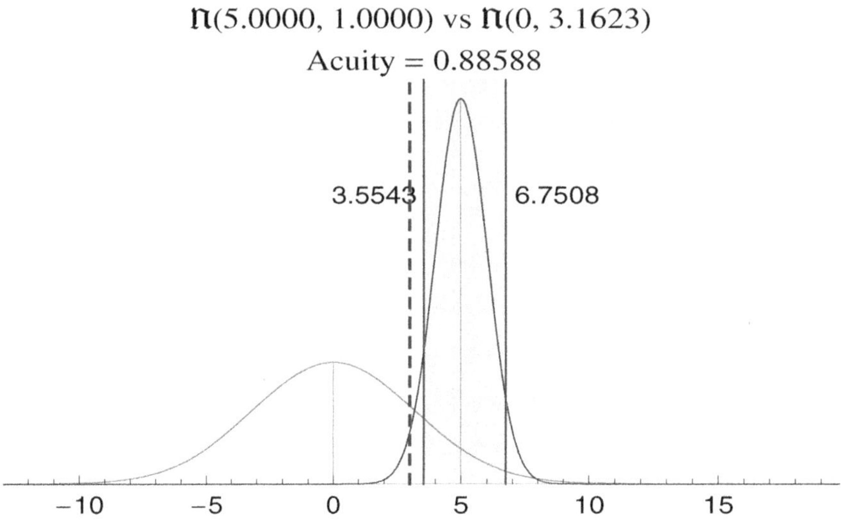

FIGURE 10.4. Maximum Acuity Decision Rule for Two Normal Distributions: $\mathfrak{N}_1(5,1)$ & $\mathfrak{N}_2(0,?)$ & Sample = $\mathfrak{N}(3,1)$

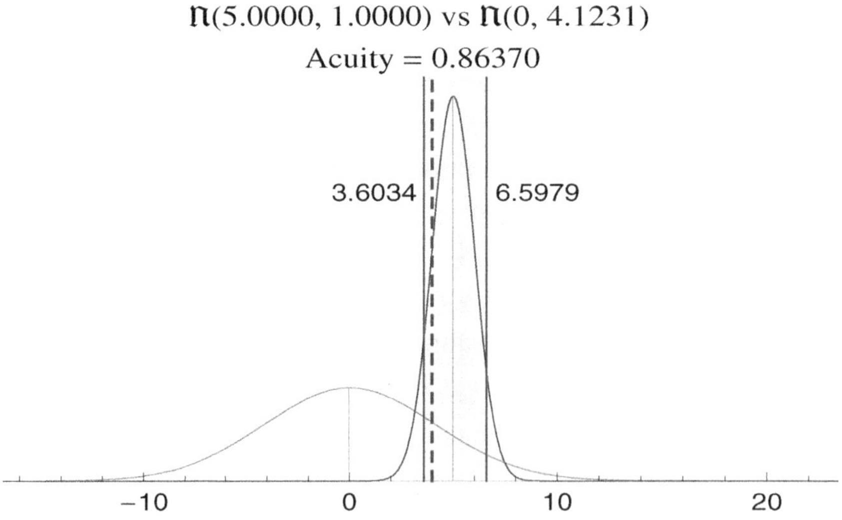

FIGURE 10.5. Maximum Acuity Decision Rule for Two Normal Distributions: $\mathfrak{N}_1(5,1)$ & $\mathfrak{N}_2(0,?)$ & Sample = $\mathfrak{N}(4,1)$

In Figure 10.6, $m = \mu_2$. At this point, the acuity reaches a minmum. The decision rule produces two cuts surrounding μ_1. m is inside the band so the decision goes to $\mathfrak{N}(\mu_2, \sigma_2)$. In contrast to Figure 10.1 the acuity is only .85910. Here there is a 5 sigma separation measured by the standard of the right-most distribution, but not the other. This is because the separation between μ_1 and μ enlarges the best estimate of σ_1.

This persists until somewhere between $m = 6$ in Figure 10.7 and $m = 7$ in Figure 10.8.

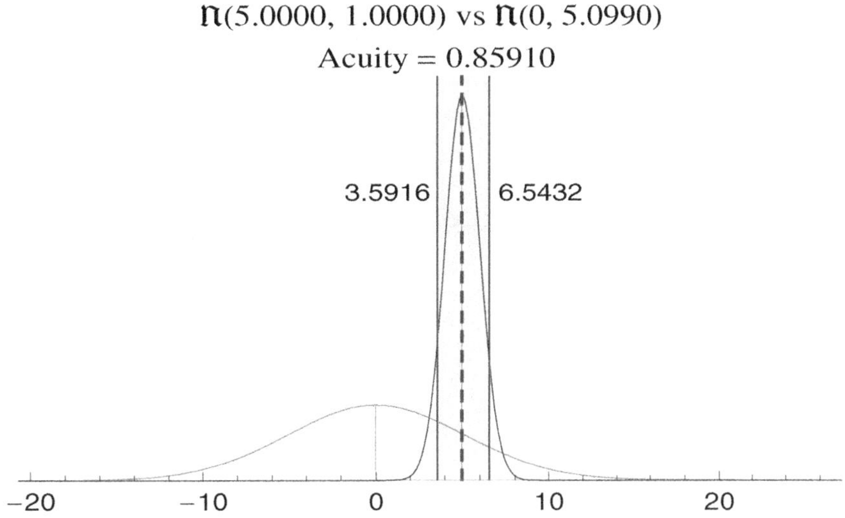

FIGURE 10.6. Maximum Acuity Decision Rule for Two Normal Distributions: $\mathfrak{N}_1(5,1)$ & $\mathfrak{N}_2(0,?)$ & Sample = $\mathfrak{N}(5,1)$

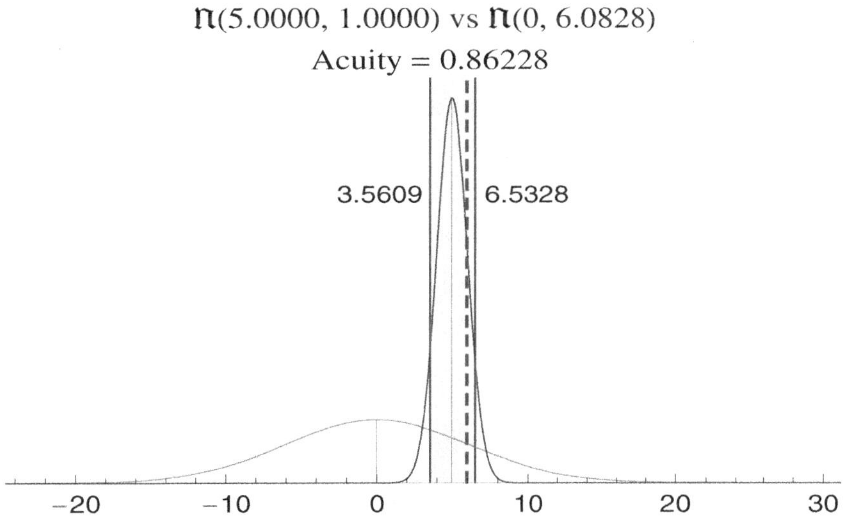

FIGURE 10.7. Maximum Acuity Decision Rule for Two Normal Distributions: $\mathfrak{N}_1(5,1)$ & $\mathfrak{N}_2(0,?)$ & Sample = $\mathfrak{N}(6,1)$

Figure 10.8 may seem anomalous. Once again m is outside the cuts that surround μ_1 so that the decision goes to $\mathfrak{N}_2(0,?)$. This is simply an unexpected consequence of a decision rule that maximizes the probability of choosing correctly.

The acuity gradually increases because as $(\mu_2 - m)^2$ tugs the value of σ_2 higher, the cuts move away from μ_1.

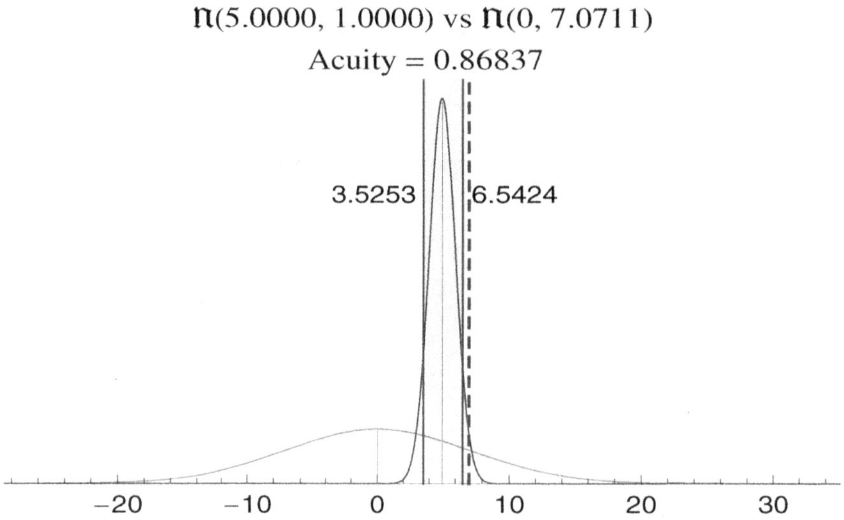

FIGURE 10.8. Maximum Acuity Decision Rule for Two Normal Distributions: $\mathfrak{N}_1(5,1)$ & $\mathfrak{N}_2(0,?)$ & Sample = $\mathfrak{N}(7,1)$

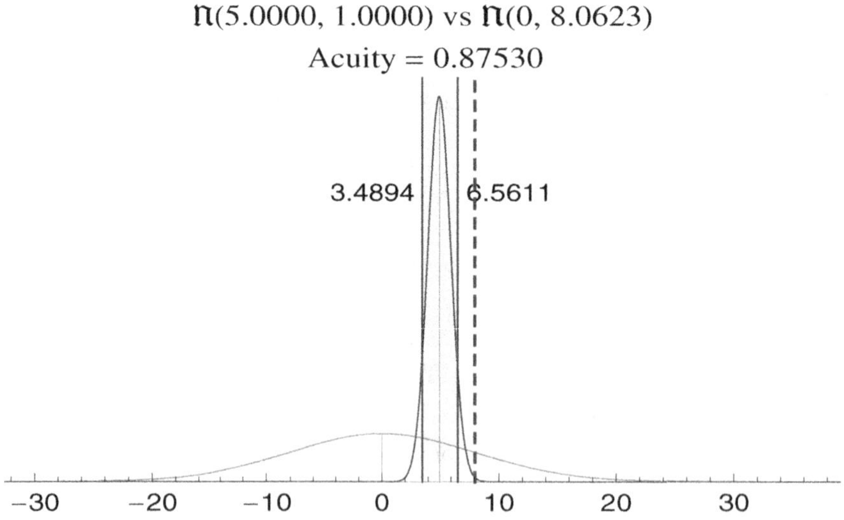

FIGURE 10.9. Maximum Acuity Decision Rule for Two Normal Distributions: $\mathfrak{N}_1(5,1)$ & $\mathfrak{N}_2(0,?)$ & Sample = $\mathfrak{N}(8,1)$

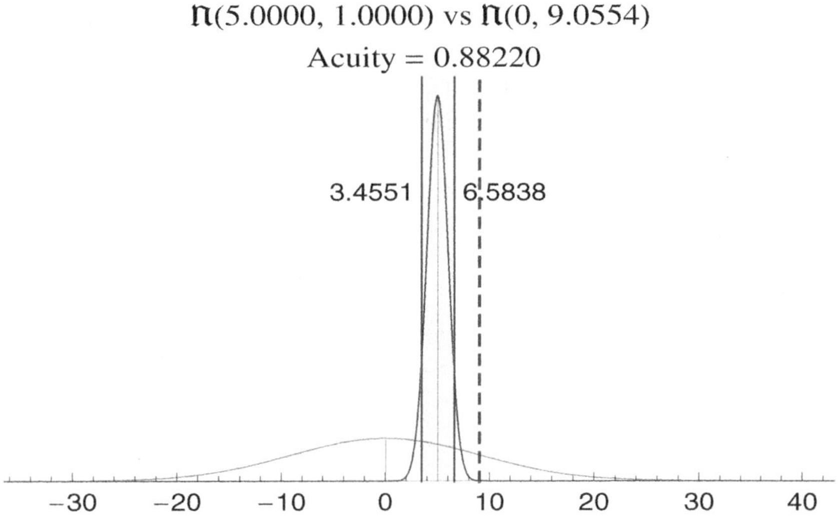

FIGURE 10.10. Maximum Acuity Decision Rule for Two Normal Distributions: $\mathfrak{N}_1(5,1)$ & $\mathfrak{N}_2(0,?)$ & Sample = $\mathfrak{N}(9,1)$

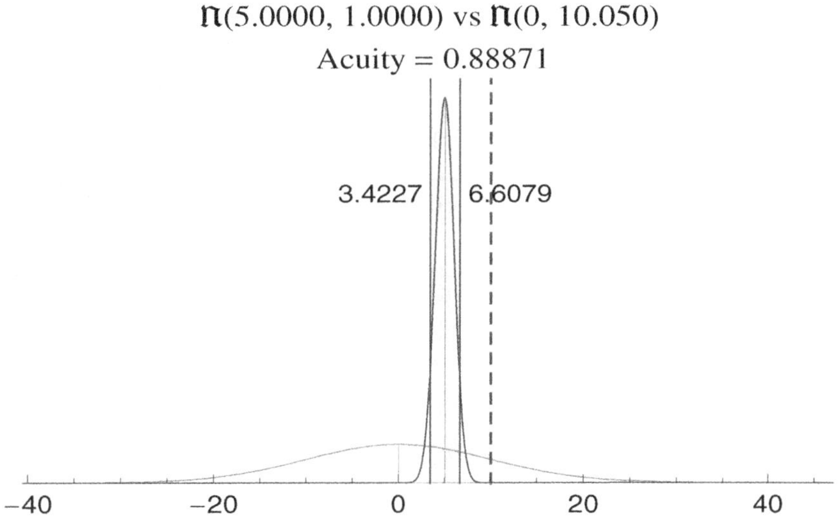

FIGURE 10.11. Maximum Acuity Decision Rule for Two Normal Distributions: $\mathfrak{N}_1(5,1)$ & $\mathfrak{N}_2(0,?)$ & Sample = $\mathfrak{N}(10,1)$

Figure 10.12 shows the three-dimensional profile of the effect of sample mean and standard deviation. Acuity is symmetric around μ_1. There is a semicircular shoulder on the graph where:

$$\left(\frac{m-\mu_1}{\sigma_2}\right)^2 + \left(\frac{s}{\sigma_2}\right)^2 = \left(\frac{\mu_2-\mu_1}{\sigma_2}\right)^2$$

Inside this boundary, where $m - \mu_1$ and s are small, the acuity of the decision rule approaches 1 as the sum of their squares approaches zero.

The value of σ_1 depends on the square root of the sum of the terms to the left in the equation above. It matters not whether $m - \mu_1$ is positive or negative; the decision rule and acuity are the same.

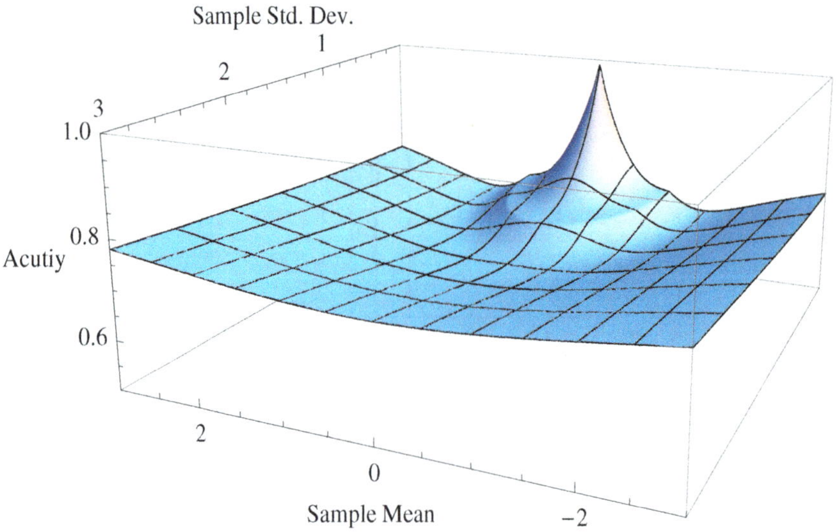

FIGURE 10.12. Acuity for Two Normal Distributions: $\mathfrak{N}_1(0, ?)$, $\mathfrak{N}_2(1, 1)$ & Sample $= \mathfrak{N}(m, s)$

That one of the two normal distributions is fixed in both mean and standard deviation greatly effects the consequences of a sample on the maximum acuity decision rule. A sample near the normal distribution that specified only by mean is has greater effect than a sample near the normal distribution that is specified by both. The sample effects only the standard deviation of the latter.

A sample with mean, m, close to the mean of the hypothesis with unspecified standard deviation and small sample standard deviation, s, can essentially refute the hypothesis that is fully specified. If the sample mean is near mean of the distribution with specified standard deviation, $(m - \mu_1)^2$ buffers the effect of that nearness on acuity. No matter how small s is, if m is not near μ_1, there will be no sharp rise towards certainty.

CHAPTER 11

$\mathfrak{N}(\mu_1, ?)$ or $\mathfrak{N}(\mu_2, ?)$ hypothesized, Sample = $\mathfrak{N}(m, s)$

In the following examples two normal distributions with means, μ_1 & μ_2 but unknown standard deviations are hypothesized. A sample is taken. Applying the the same logic as in the previous chapter, but doing it twice.

$$\sigma_1 = \sqrt{s^2 + (\mu_1 - m)^2}$$
$$\sigma_2 = \sqrt{s^2 + (\mu_2 - m)^2}$$

The maximum acuity decision rule and resultant acuity is calculated for $\mathfrak{N}(\mu_1, \sigma_1)$ and $\mathfrak{N}(\mu_2, \sigma_2)$. The decision goes to the distribution whose mean is closest to the sample mean, but the acuity varies greatly depending on the values of $m, s, \mu_1,$ and μ_2.

	Unequal μ, Both σ Unknown The Decision Rule, $\boldsymbol{d}$, Is Determined by Two Cuts at $\boldsymbol{a}$ & $\boldsymbol{b}$	
	$\boldsymbol{a < m > b}$	$\boldsymbol{m < a}$ or $\boldsymbol{m > b}$
$\mathfrak{h}_1$: Whose μ is between the cuts	$\boldsymbol{\mathcal{A}_d}$	$1 - \boldsymbol{\mathcal{A}_d}$
$\mathfrak{h}_2$: Whose μ is outside the cuts	$1 - \boldsymbol{\mathcal{A}_d}$	$\boldsymbol{\mathcal{A}_d}$
	say: "$\mathfrak{h}_1$ is true"	say: "$\mathfrak{h}_2$ is true"
	The Decision: Acuity = $\boldsymbol{\mathcal{A}_d}$	

TABLE 11.1. Unequal $\boldsymbol{\mu}$, Both $\boldsymbol{\sigma}$ Unknown

Figure 11.1 is deciding between the truth of a normal distribution with mean = 0 and a normal distribution with mean = 5. The sample has a mean = 0 and standard deviation = 1. Because $m = \mu_1$, $\sigma_1 = s$. $(\mu_1 - m)^2 = 0$ and does not affect the result.

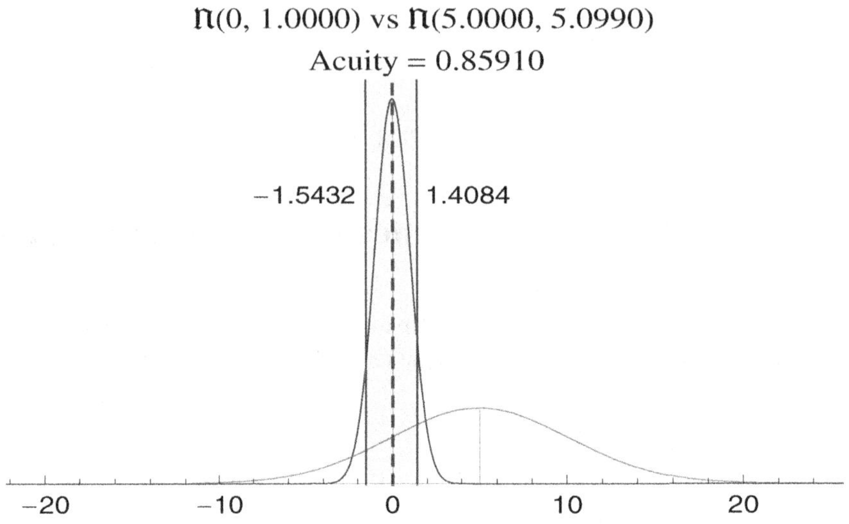

FIGURE 11.1. Maximum Acuity Decision Rule for Two Normal Distributions: $\mathfrak{N}_1(0, ?)$ & $\mathfrak{N}_2(5, ?)$ & Sample = $\mathfrak{N}(0, 1)$

From here the acuity declines until just before the sample mean is midway between μ_1 and μ_2. The decision always goes to $\mathfrak{N}(\mu_1, ?)$.

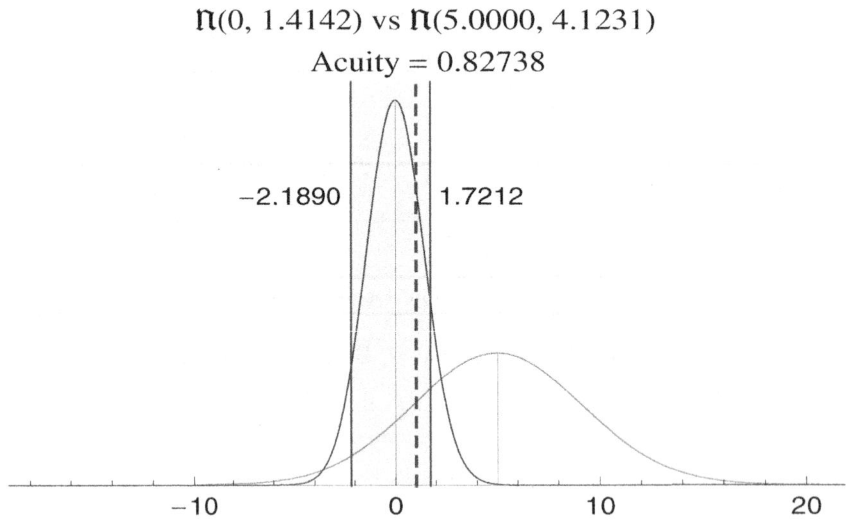

FIGURE 11.2. Maximum Acuity Decision Rule for Two Normal Distributions: $\mathfrak{N}_1(0, ?)$ & $\mathfrak{N}_2(5, ?)$ & Sample = $\mathfrak{N}(1, 1)$

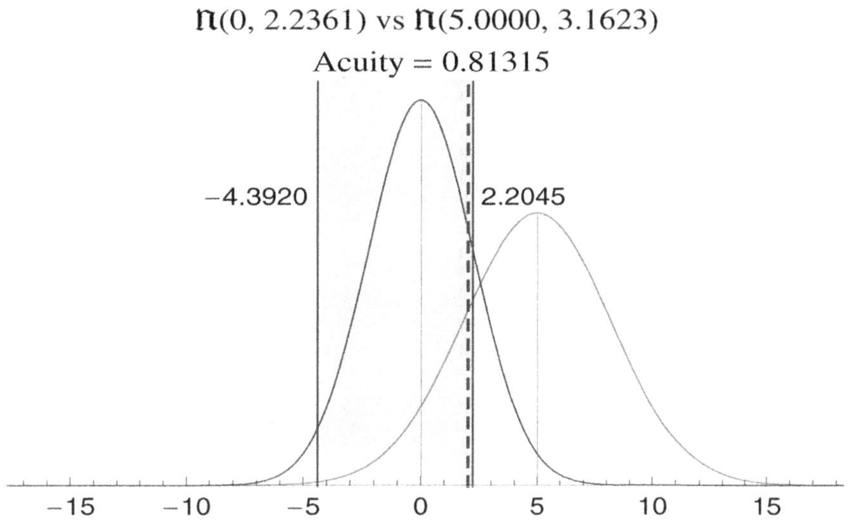

FIGURE 11.3. Maximum Acuity Decision Rule for Two Normal Distributions: $\mathfrak{N}_1(0, ?)$ & $\mathfrak{N}_2(5, ?)$ & Sample = $\mathfrak{N}(2, 1)$

Figure 11.4 is an example of the rare occurrence of sample evidence where no decision is possible. The decision is between normal distribution with mean = 0 and a normal distribution with mean = 2. The sample evidence has a normal distribution with mean = 1 and standard deviation = 1. The mean of the sample is exactly equal to dividing line between the two distribution—no decision.

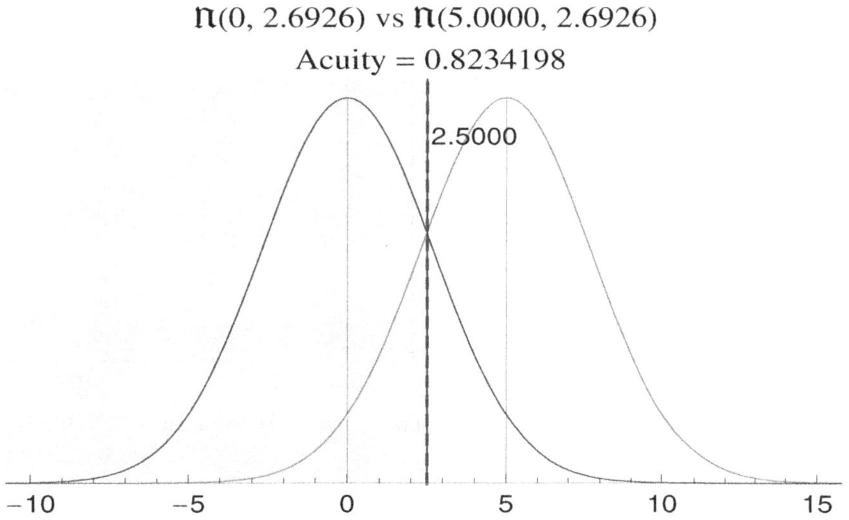

FIGURE 11.4. Maximum Acuity Decision Rule for Two Normal Distributions: $\mathfrak{N}_1(0, ?)$ & $\mathfrak{N}_2(5, ?)$ & Sample = $\mathfrak{N}(2.5, 1)$

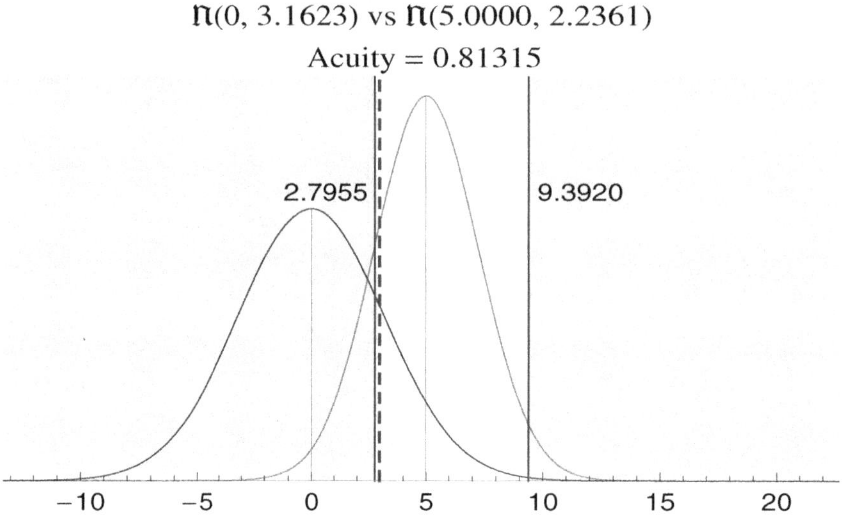

FIGURE 11.5. Maximum Acuity Decision Rule for Two Normal Distributions: $\mathfrak{N}_1(0, ?)$ & $\mathfrak{N}_2(5, ?)$ & Sample = $\mathfrak{N}(3, 1)$

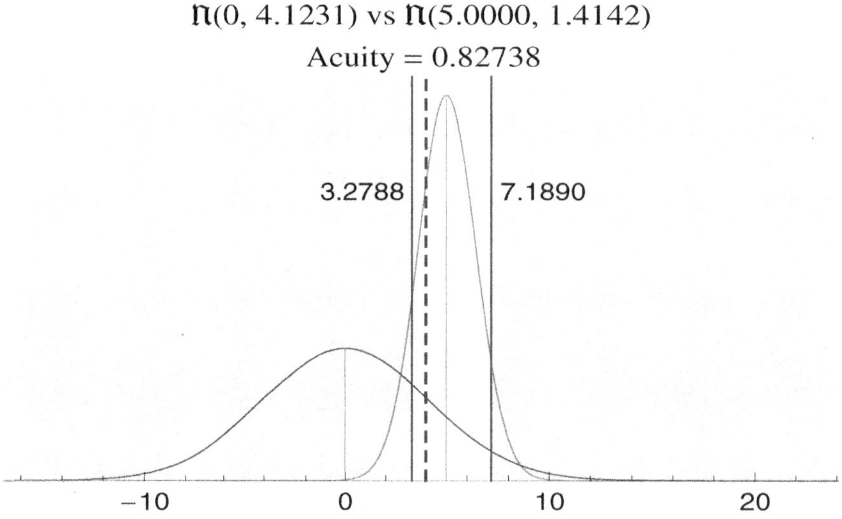

FIGURE 11.6. Maximum Acuity Decision Rule for Two Normal Distributions: $\mathfrak{N}_1(0, ?)$ & $\mathfrak{N}_2(5, ?)$ & Sample = $\mathfrak{N}(4, 1)$

In Figure 11.7, again m equals one of the two hypothesized means. $m = \mu_2$ so $\sigma_2 = s$ and the aquity reaches a peak and then drifts lower as m moves to the left.

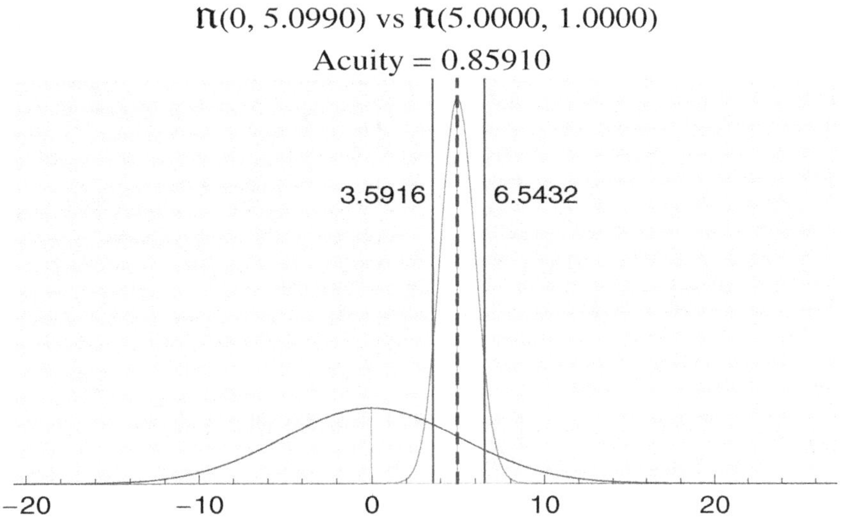

FIGURE 11.7. Maximum Acuity Decision Rule for Two Normal Distributions: $\mathfrak{N}_1(0, ?)$ & $\mathfrak{N}_2(5, ?)$ & Sample = $\mathfrak{N}(5, 1)$

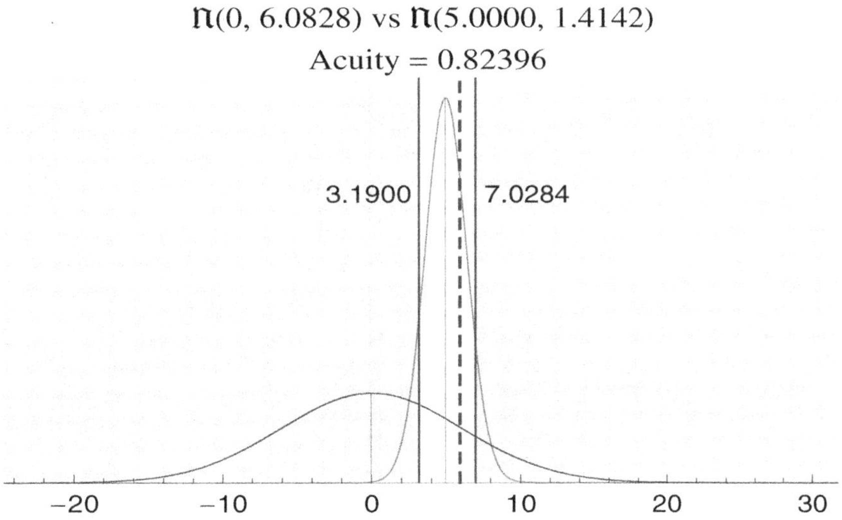

FIGURE 11.8. Maximum Acuity Decision Rule for Two Normal Distributions: $\mathfrak{N}_1(0, ?)$ & $\mathfrak{N}_2(5, ?)$ & Sample = $\mathfrak{N}(6, 1)$

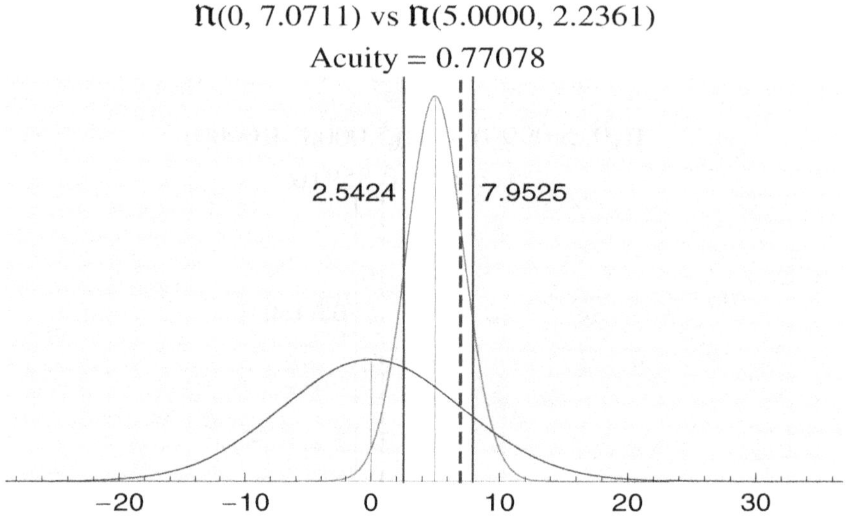

FIGURE 11.9. Maximum Acuity Decision Rule for Two Normal Distributions: $\mathfrak{N}_1(0, ?)$ & $\mathfrak{N}_2(5, ?)$ & Sample = $\mathfrak{N}(7, 1)$

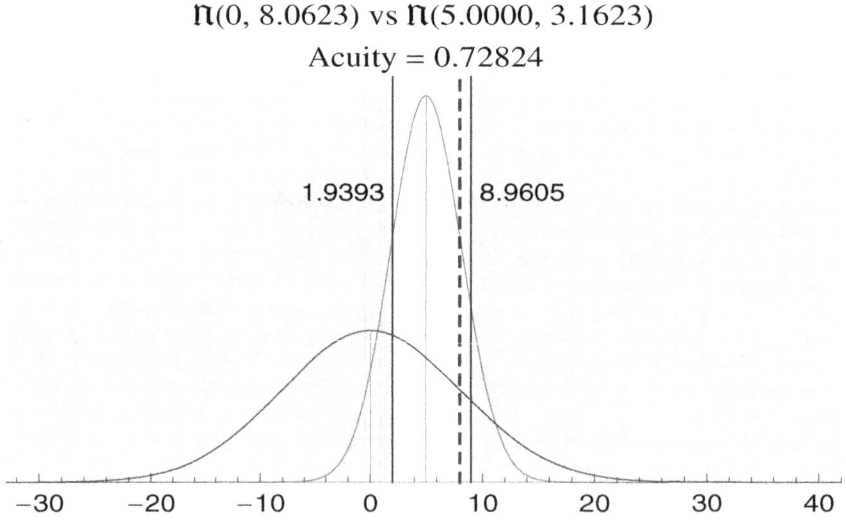

FIGURE 11.10. Maximum Acuity Decision Rule for Two Normal Distributions: $\mathfrak{N}_1(0, ?)$ & $\mathfrak{N}_2(5, ?)$ & Sample = $\mathfrak{N}(8, 1)$

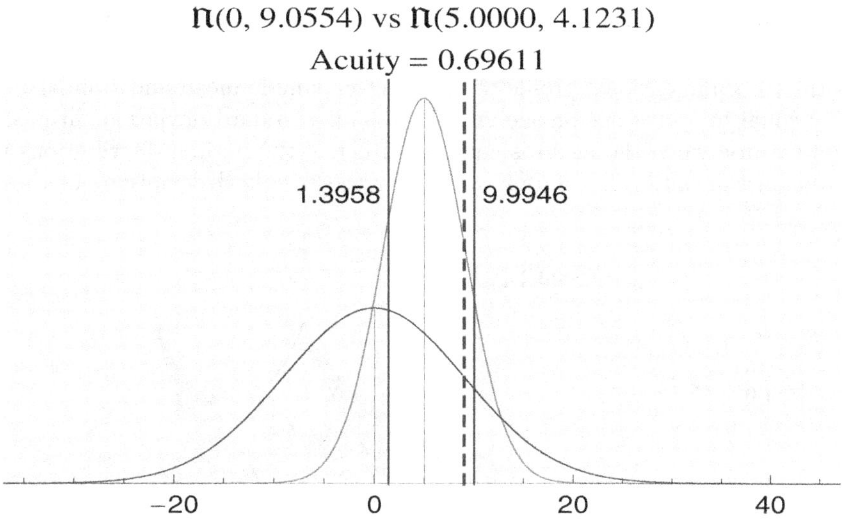

FIGURE 11.11. Maximum Acuity Decision Rule for Two Normal Distributions: $\mathfrak{N}_1(0, ?)$ & $\mathfrak{N}_2(5, ?)$ & Sample = $\mathfrak{N}(9, 1)$

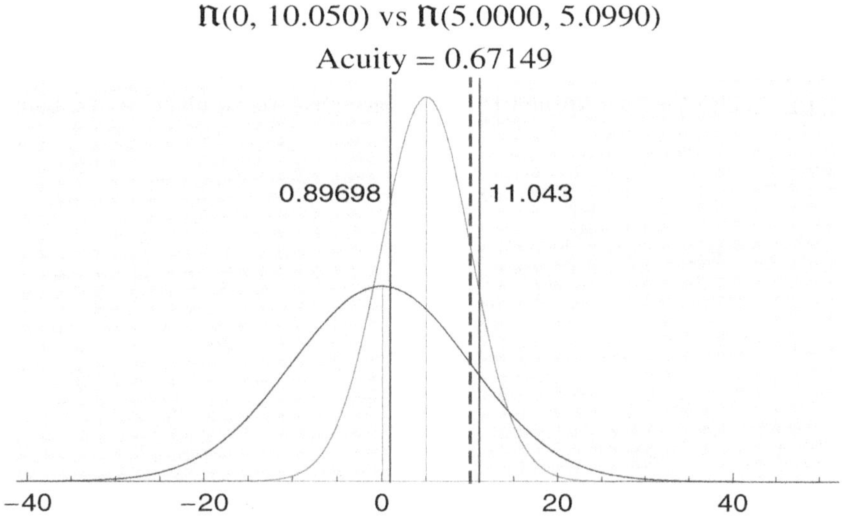

FIGURE 11.12. Maximum Acuity Decision Rule for Two Normal Distributions: $\mathfrak{N}_1(0, ?)$ & $\mathfrak{N}_2(5, ?)$ & Sample = $\mathfrak{N}(10, 1)$

Because $\mathfrak{N}(m, s)$ pulls on both hypothesized distributions, m always falls within the cuts that surround the nearer distribution, but as $\mathfrak{N}(m, s)$ moves away from the midpoint between the hypothesized distributions, the acuity declines toward $\frac{1}{2}$.

This is quite different from Chapter 10 where m moved *past* the cuts so that in the extreme, the decision went to the distant hypothesis.

Figure 11.13 shows the three-dimensional profile of the effect of sample mean and standard deviation. In particular, when the sample mean is equal to minus one or one and the sample standard deviation shrinks toward zero, the acuity rises toward one. This is because when the mean is equal to either of the normal distributions with unknown standard deviation, the sample standard deviation is the only driver for the standard deviations of those normal distributions, $\mathfrak{N}(-1, ?)$ and $\mathfrak{N}(1, ?)$.

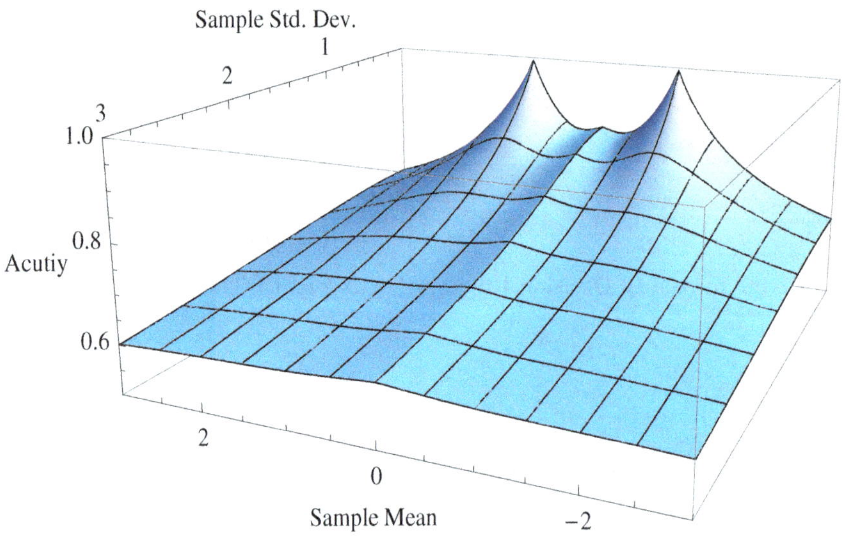

FIGURE 11.13. Acuity for Two Normal Distributions: $\mathfrak{N}_1(-1, ?)$, $\mathfrak{N}_2(1, ?)$, & Sample = (m, s)

CHAPTER 12

$\mathfrak{N}(?,?)$ or $\mathfrak{N}(\mu_2,?)$ hypothesized, Sample $= \mathfrak{N}(m,s)$

This is the situation that practitioners in the soft sciences find themselves. The first alternative is completely unspecified, with unknown mean an unknown standard deviation. The second alternative is a null hypothesis about which all that is known is the value of its mean . To fill in the missing data, the standard deviation for one distribution and the mean and standard deviation for the other we rely on Postulate III, page 28.

In the first instance, $\mathfrak{N}(?,?) = \mathfrak{N}(m,s)$—reality is like the sample. We simplify by assuming that $s = 1$. In the second instance, we further simplify by assuming that $\mu_1 = 0$. Therefore,

$$\mathfrak{N}(\mu_2,?) = \mathfrak{N}(0, \sqrt{s^2 + (\mu_2 - m)^2}) = \mathfrak{N}(0, \sqrt{1 + m^2}).$$

This looks like the iconic t-test model for determining the significance of the difference between a group receiving a treatment and one receiving no treatment. One finds the mean and standard deviation of the difference between the groups and calculates the significance of the difference versus zero difference.

We can answer the question that significance tests cannot: given the sample as evidence and given that either the null hypothesis or the sample are reality, what is the probability that reality is like the null hypothesis versus the probability that reality is like the sample?

	One μ Unknown & Both σ Unknown The Decision Rule, $\boldsymbol{d}$, Is Determined by Two Cuts at $\boldsymbol{a}$ & $\boldsymbol{b}$	
	$\boldsymbol{a} < \boldsymbol{m} > \boldsymbol{b}$	$\boldsymbol{m} < \boldsymbol{a}$ or $\boldsymbol{m} > \boldsymbol{b}$
$\mathfrak{h}_1: \ \mathfrak{N}(m,s)$	$\boldsymbol{\mathcal{A}_d}$	$1 - \boldsymbol{\mathcal{A}_d}$
$\mathfrak{h}_2: \ \mathfrak{N}(0, \sqrt{s_m^2 + m^2})$	$1 - \boldsymbol{\mathcal{A}_d}$	$\boldsymbol{\mathcal{A}_d}$
	say: "$\mathfrak{h}_1$ is true"	say: "$\mathfrak{h}_2$ is true"
	The Decision: Acuity $= \boldsymbol{\mathcal{A}_d}$	

TABLE 12.1. One $\boldsymbol{\mu}$ Unknown & Both $\boldsymbol{\sigma}$ Unknown

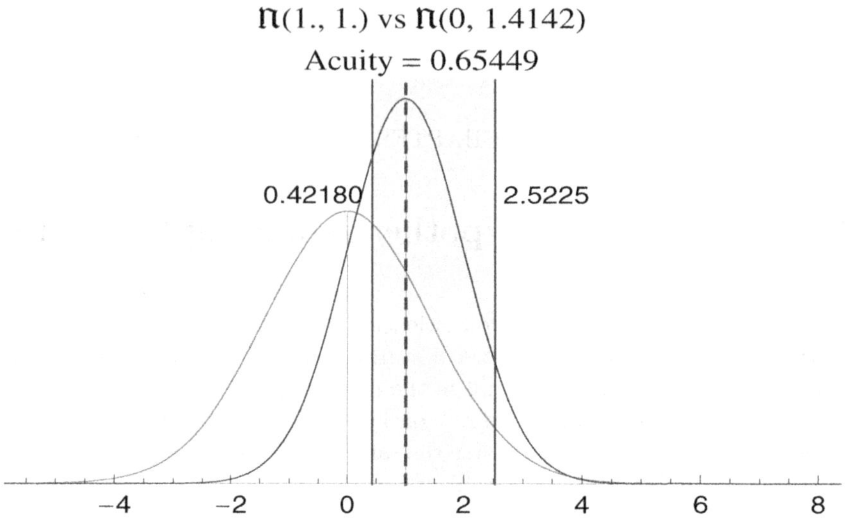

FIGURE 12.1. Acuity for A Sample = $\mathfrak{N}(1,1)$ & Two Normal Distributions: $\mathfrak{N}_1(?,?)$ vs $\mathfrak{N}_2(0,?)$

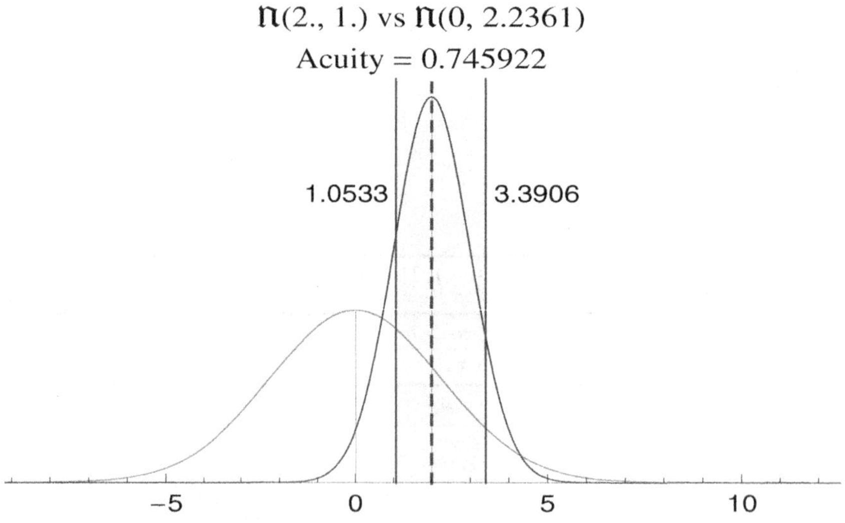

FIGURE 12.2. Acuity for A Sample = $\mathfrak{N}(2,1)$ & Two Normal Distributions: $\mathfrak{N}_1(?,?)$ vs $\mathfrak{N}_2(0,?)$

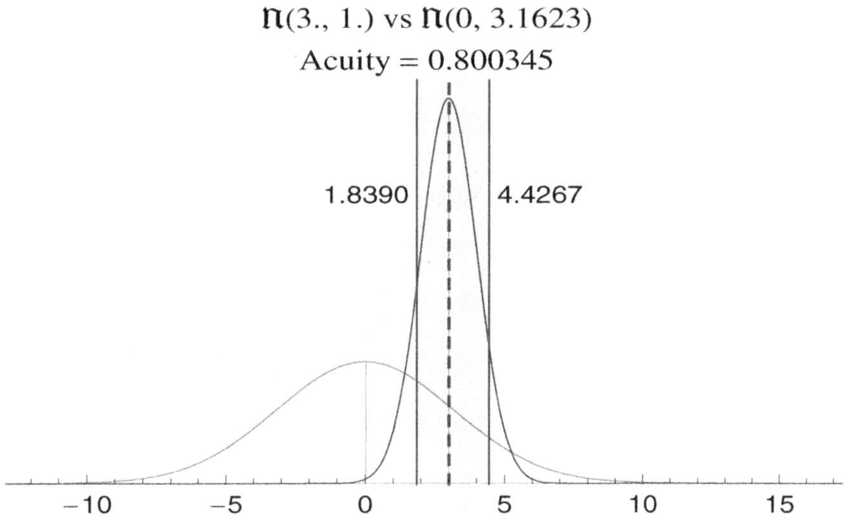

FIGURE 12.3. Acuity for A Sample = $\mathfrak{N}(3,1)$ & Two Normal Distributions: $\mathfrak{N}_1(?,?)$ vs $\mathfrak{N}_2(0,?)$

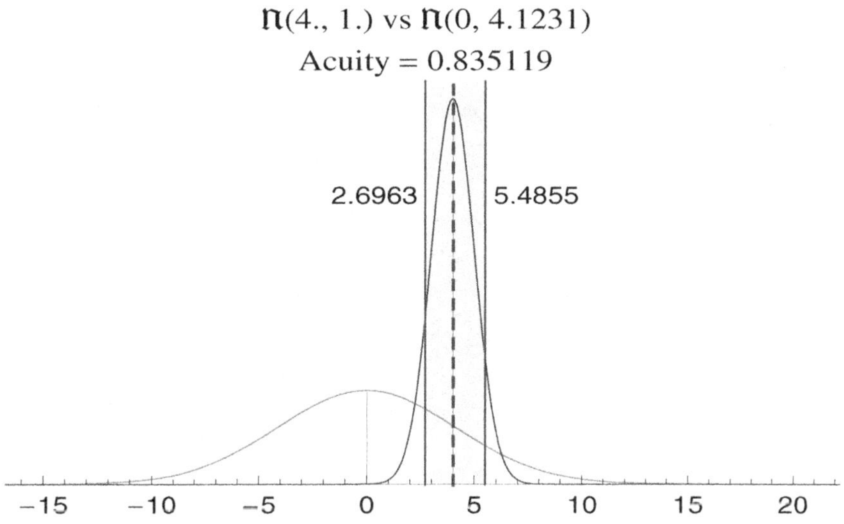

FIGURE 12.4. Acuity for A Sample = $\mathfrak{N}(4,1)$ & Two Normal Distributions: $\mathfrak{N}_1(?,?)$ vs $\mathfrak{N}_2(0,?)$

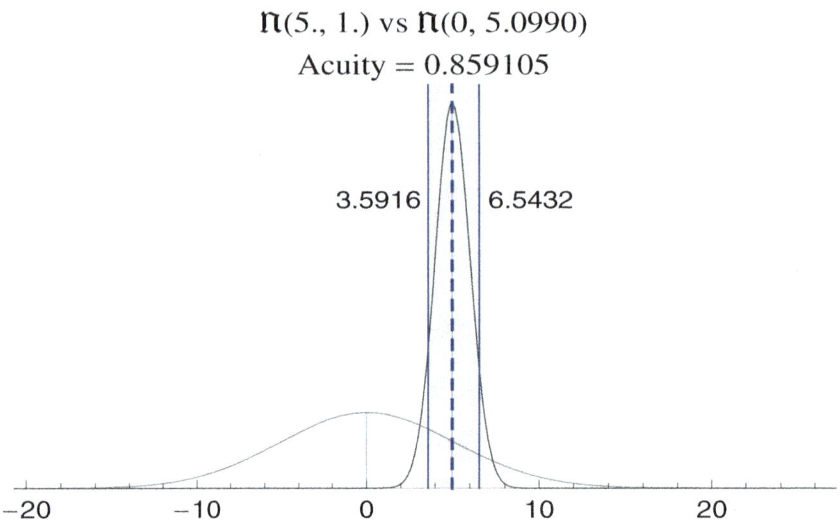

FIGURE 12.5. Acuity for A Sample = $\mathfrak{N}(5,1)$ & Two Normal Distributions: $\mathfrak{N}_1(?,?)$ vs $\mathfrak{N}_2(0,?)$

The decision always goes to the sample, $\mathfrak{N}(m,s)$, but if s is not small and m is not far from μ_2, the acuity is lackluster.

The next graph is a three dimensional plot of acuity versus the mean and standard deviation of the sample. The only way to get high acuity is for

$$\frac{|\mu_2 - m|}{s}$$

to be large.

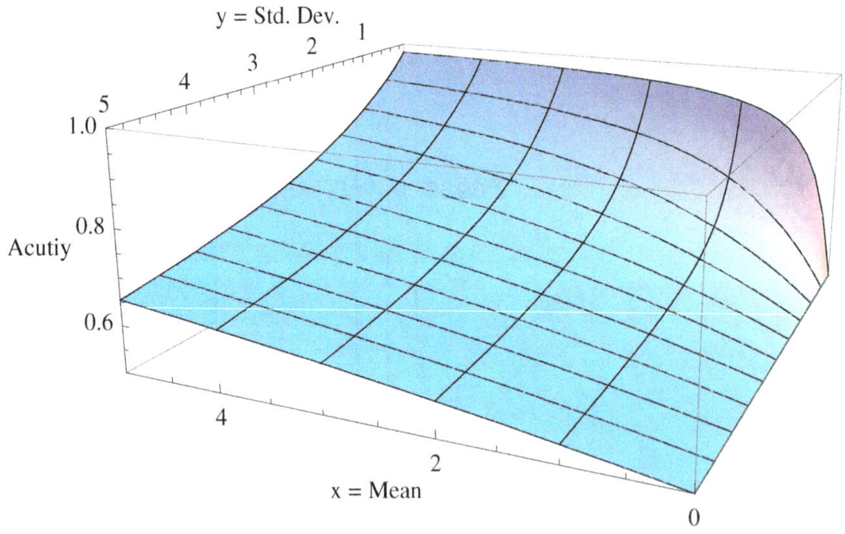

FIGURE 12.6. Maximum Acuity for Null Normal Distribution, $\mathfrak{N}(0,?)$ and Sample $\mathfrak{N}(x,y)$

In the soft sciences, this happy confluence of circumstances is the exception. It is common for a result to be "statistically significant," but rare for the best decision rule to have high acuity.

Next we will directly examine the interaction of acuity with significance testing and how acuity differs in the hard and soft sciences. The results are not good for the home team.

CHAPTER 13

Significance Tests

Acuity of t-tests

Before computers, one would then have gone to a table to the t-distribution, looked in the row for $n-1$ degrees of freedom, and proclaimed the result "not significant" or "significant at the ... level!" The result presents a slight problem because in the computer age one is quickly off the table. My old table stops at 3.373, equivalent to being significant at the .0005 level with 120 degrees of freedom.

In the past, computational necessity and professional custom joined forces to limit the cutoff points to less extreme levels. The usual choices are significance at the 5%, 1%, and 0.1% levels. To quote a standard textbook:

> Generally, values of α greater than .05 are not considered to be *statistically significant.* A significance level of 5% ($P = 0.05$) corresponds to one type I error in 20 trials, a level of 1% ($P = 0.01$) of one error in 100 trials. Significance of levels of less than 1% ($P = 0.01$) are nearly always adjudged significant; those between 5% and 1% may be considered significant at the discretion of the investigator. [**46**][p. 164]

The authors might have added that $P = 0.001$ is often considered "highly significant." Later they continue explaining the social conventions of their field:

> A convention often encountered is the use of asterisks after the computed value of the significance test, $t_s = 2.86^{**}$. The symbols generally represent the following probability ranges:
>
> $$* = 0.05 \geq P > 0.01 \quad ** = 0.01 \geq P > 0.001 \quad *** = P \leq 0.001$$
>
> [**46**][p. 172]

It is generally regarded as a lapse of good taste to claim significance at a level higher than 0.001. Extremely high levels of certainty on this question are regarded as implausible. Paying too close attention to the percentile actually achieved is considered skating on thin ice. If one does not pay close attention, one might mistake the significance level for the probability that the tested hypothesis is true.[1]

The following graphs show the acuity that is implicit when a sample is significant at a certain level of alpha in a one-tailed t-test. Again we make the simplifying assumption that the number of data points in the sample is large enough for the normal distribution to be an adequate approximation of the t-distribution. In the graphs, we have rearranged the distributions so that α is that part of the sample distribution that is *less* than zero. Ordinary usage would have the distribution centered on zero with the sample mean in the right tail, but we need to show the distribution of the sample mean where it is located.

At the level of one in 10 odds, results are rarely published, but researchers may feel that they're on to something and want to enlarge the sample.

---

[1]It is commonplace to hear, "There is only a 1% probability that these results were caused by chance alone." To me this seems even closer to relabeling significance level as probability of truth of the hypothesis proposed because it is not entirely clear, by the language alone, whether "the results" refers to rejecting the null hypothesis or the *reality* of the sample.

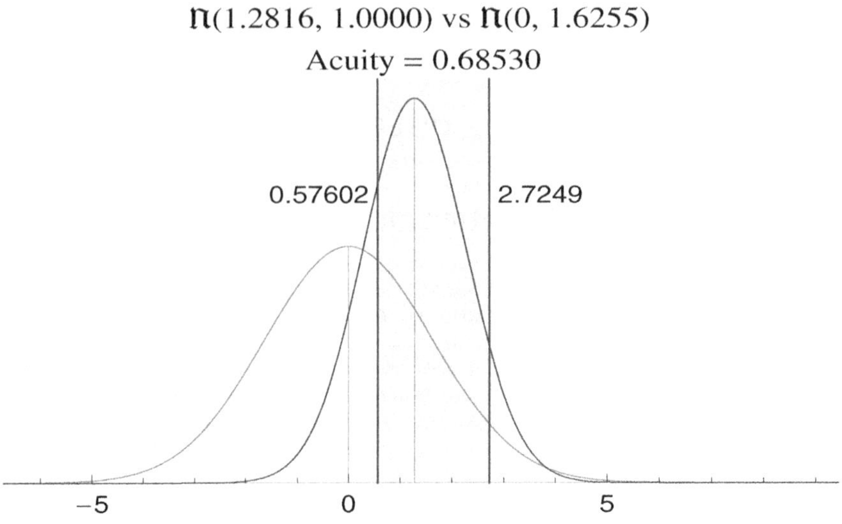

FIGURE 13.1. The Acuity of a one-tailed t-test with $\alpha = .10$

One in 20 odds is the level at which results in the soft sciences begin to be published. It is also the level at which the results of significance tests begin to be overblown. If one is to choose between the null hypothesis or a world like the sample, the odds of choosing correctly are not 19 to 1, but rather about 2.6 to 1.

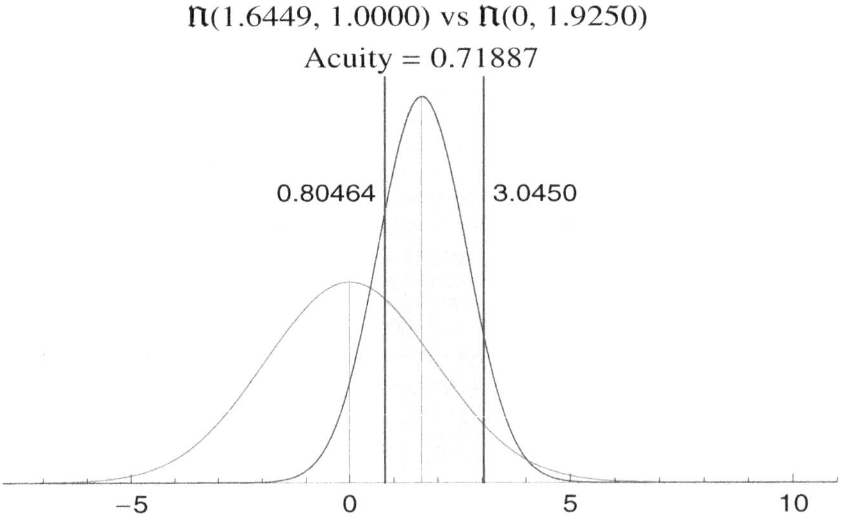

FIGURE 13.2. The Acuity of a one-tailed t-test with $\alpha = .05$

Rejecting the null hypothesis at the level of 99 to one is beginning to be a level of clear certainty, if one were to believe the significance tests. The actual probability of choosing correctly is about 3.3 to one.

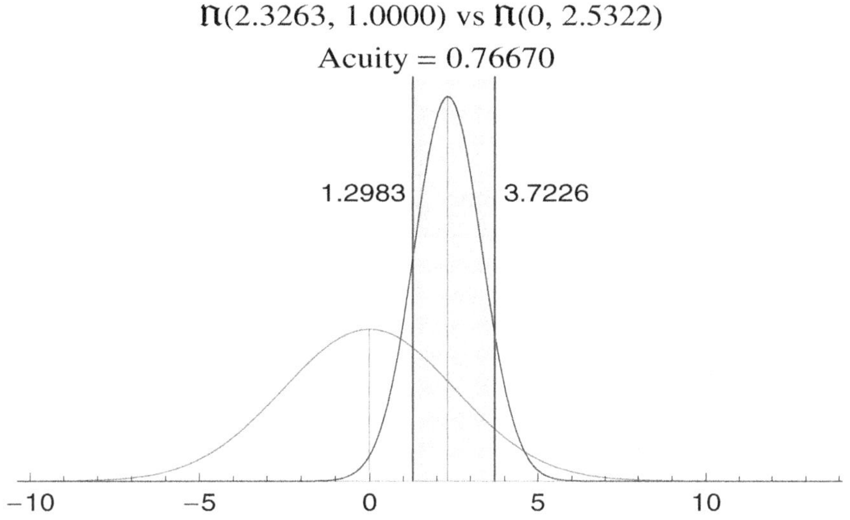

FIGURE 13.3. The Acuity of a one-tailed t-test with $\alpha = .01$

As shown in Figure 13.4, this is also the place where we begin to run out of asterisks and begin to report the P value or state that the P value is less than some amount. This is the place where it begins to be unseemly to report the actual results of the t-test by quoting the actual numbers. But if the question is whether the null hypothesis is true or the world is like the sample the probability of being incorrect is not one in 1000, but about the same as coming out unscathed playing Russian roulette with a five shot revolver, 4.1 to 1.

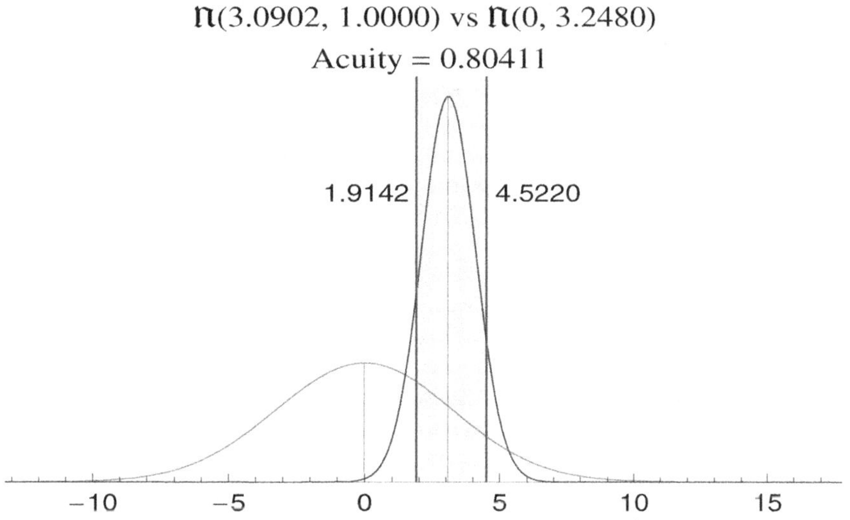

FIGURE 13.4. The Acuity of a one-tailed t-test with $\alpha = .001$

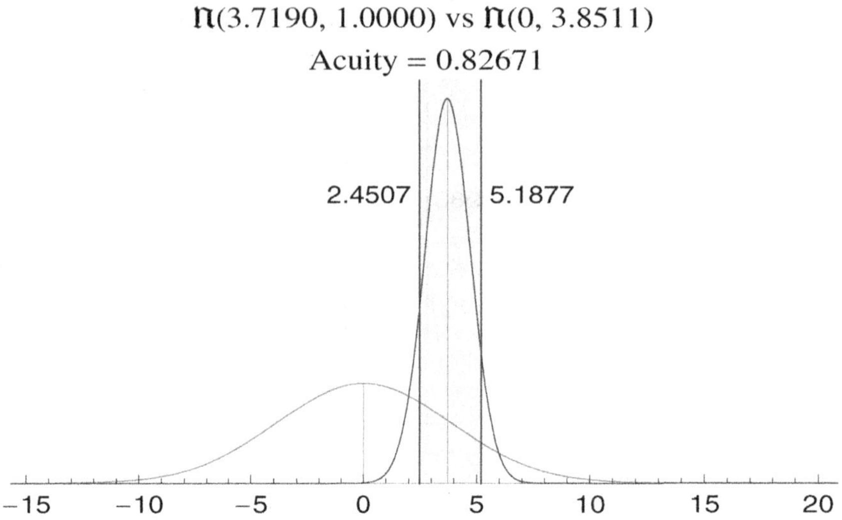

FIGURE 13.5. The Acuity of a one-tailed t-test with $\alpha = .0001$

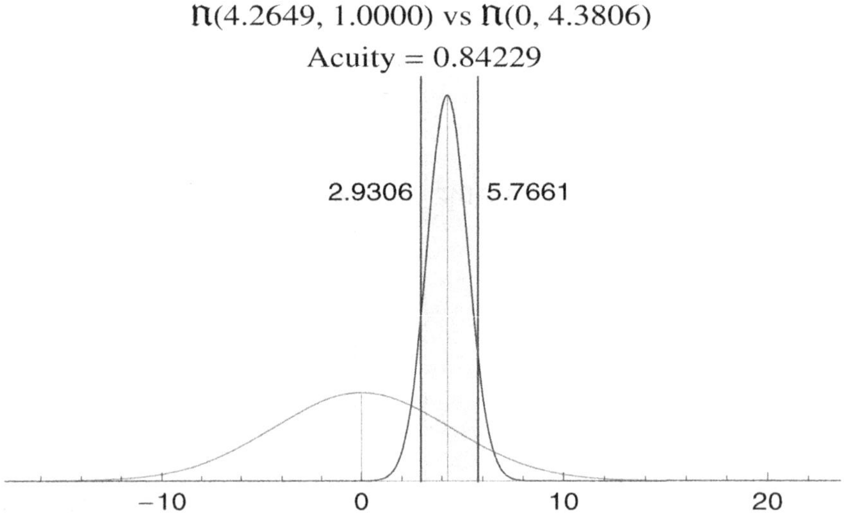

FIGURE 13.6. The Acuity of a one-tailed t-test with $\alpha = .00001$

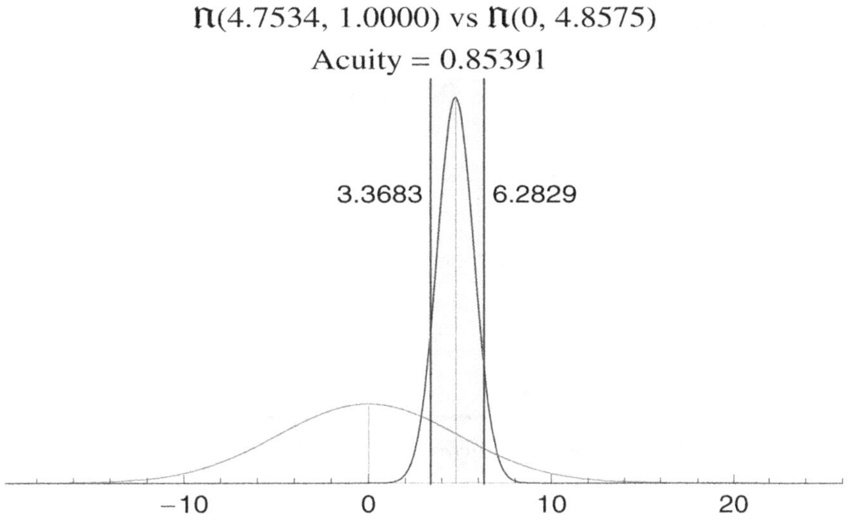

FIGURE 13.7. The Acuity of a one-tailed t-test with $\alpha = .000001$

The results can only be seen as disappointing. Using the best decision rule, the odds of correctly rejecting the null hypothesis in favor the sample, if either the null hypothesis or the sample represents the true state of the world is shown in Table 13.1

α	**Odds of Choosing Correctly**
.10	2.178 to 1
.05	2.557 to 1
.01	3.286 to 1
.001	4.105 to 1
.0001	4.771 to 1
.00001	5.341 to 1
.000001	5.845 to 1

TABLE 13.1. Odds of Correctly Rejecting the Null Hypothesis

The probability incorrectly rejecting the null hypothesis in favor of the sample with:

α	**Probability of Getting It Wrong**
.10	.3147
.05	.2811
.01	.2333
.001	.1959
.0001	.1733
.00001	.1577
.000001	.1461

TABLE 13.2. Probability of Incorrectly Rejecting the Null Hypothesis

The following graph goes from $\alpha = .10$ to $\alpha = 10^{-40}$. On the left odds of 9 to 1 of the sample being drawn from the null hypothesis is the same as odds of 2.17765 to 1 that one would correctly choose the sample over the null hypothesis. On the right odds of 1 to 9, 999, 999, 999, 999, 999, 999, 999, 999, 999, 999, 999, 999, 999 of the sample being drawn from the null hypothesis is the same as odds of 14.01315 to 1 that one would correctly choose a normal distribution like the sample over the null hypothesis.

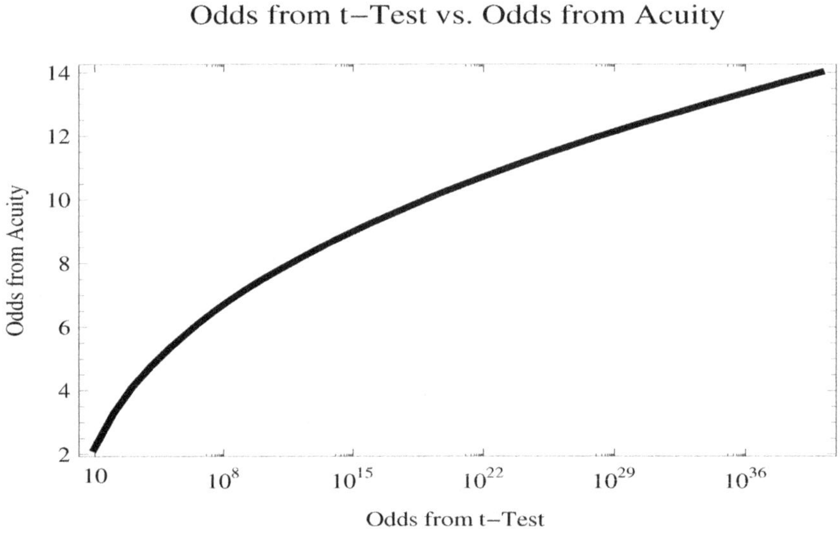

FIGURE 13.8. Odds Derived from a *t*-Test vs. Odds Derived from Acuity

What are to practical implications of this disparity? Data that can help answer that question has been collected.

Wetzels *et al.* " . . . studied all of the comparisons evaluated by a *t*-test in the Year 2007 volumes of Psychonomic Bulletin & Review (PBR) and Journal of Experimental Psychology: Learning, Memory and Cognition (JEP:LMC). This sample was comprised of 855 comparisons from 252 articles. These articles covered 2394 journal pages, and addressed many topics that are important in modern experimental psychology. Our sample suggests, on average, that an article published in PBR and JEP:LMC contains about 3.4 t tests, which amounts to one t test every 2.8 pages." [**53**] Through personal correspondence with Ruud Wetzels,[2] I received more fine-grained data. That is the basis for Table 13.3 below.

Published Statistically Significant Results						
α	≤ 0.000001	≤ 0.00001	≤ 0.0001	≤ 0.001	≤ 0.01	≤ 0.05
% of Total	16%	20%	27%	42%	69%	100%
% Untrue	$\leq 14.6\%$	$\leq 15.8\%$	$\leq 17.3\%$	$\leq 19.6\%$	$\leq 23.3\%$	$\leq 28.1\%$
Odds of Being Correct	≥ 5.845	≥ 5.341	≥ 4.771	≥ 4.105	≥ 3.286	≥ 2.557

TABLE 13.3. Of the 855, Those Found Significant at $\alpha = .05$

Looking at the column headed with $\alpha \leq 0.0001$, seventy-three percent of the published positive results, have worse odds of being true, than you have of surviving Russian roulette with a six-shooter. This is not what wants if one is trying to establish scientific truth.

Acuity vs. Significance in the Soft Sciences

The following two charts, Figures 13.9 and 13.10, show the Acuity and odds of choosing the correct hypothesis when one choice is $\mathfrak{N}(0,1)$, a normal distribution with $\mu = 0$ and $\sigma = 1$, and the alternative is $\mathfrak{N}(x,?)$, a normal distribution with $\mu = x$, the value on the x axis, and σ unknown. This is the situation usually faced by investigators in the soft sciences.

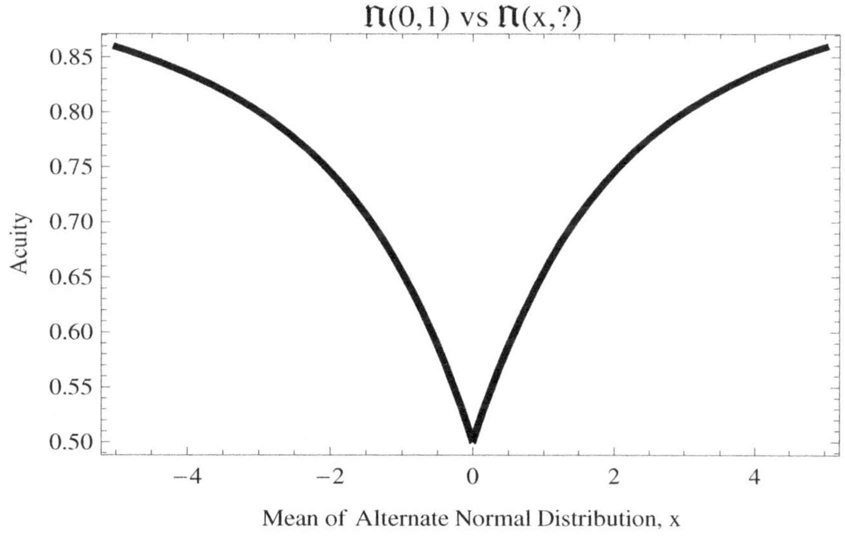

FIGURE 13.9. Acuity for $\mathfrak{N}(0,1)$ vs $\mathfrak{N}(x,?)$

[2] 2015-02-15

When the means differ by five sample standard deviations, in the soft sciences, there is only about 86% probability that the correct choice will be made between the null hypothesis and the hypothesis that the world is like the sample. This is because the mean of the sample helps mold the proper estimate of the standard deviation of the null hypothesis.

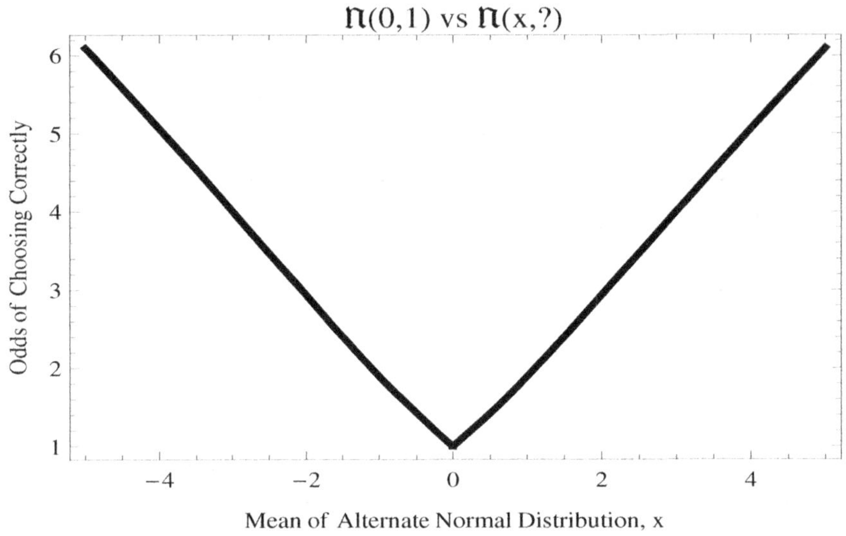

FIGURE 13.10. Odds for $\mathfrak{N}(0,1)$ vs $\mathfrak{N}(x,?)$

The odds increase more or less linearly as x moves away from 0 at about $\frac{6}{5}$ of the difference in means. This is pathetically slow.

Vague hypotheses are hard to validate. Significance testing has produced an inflated sense of certainty that is not justified and far below what is standard in the hard sciences, as will be shown in the next section.

Acuity vs. Significance in the Hard Sciences

The next two charts, Figures 13.11 and 13.12, show the Acuity, odds of choosing the correct hypothesis, and the logarithm of those odds when one choice is $\mathfrak{N}(0,1)$, a normal distribution with $\mu = 0$ and $\sigma = 1$, and the alternative is $\mathfrak{N}(x,1)$, a normal distribution with $\mu = x$, the value on the x axis, and $\sigma = 1$. This is the situation usually faced by investigators in the hard sciences where the error is in the instrumentation and is assumed to be the same for both hypotheses.

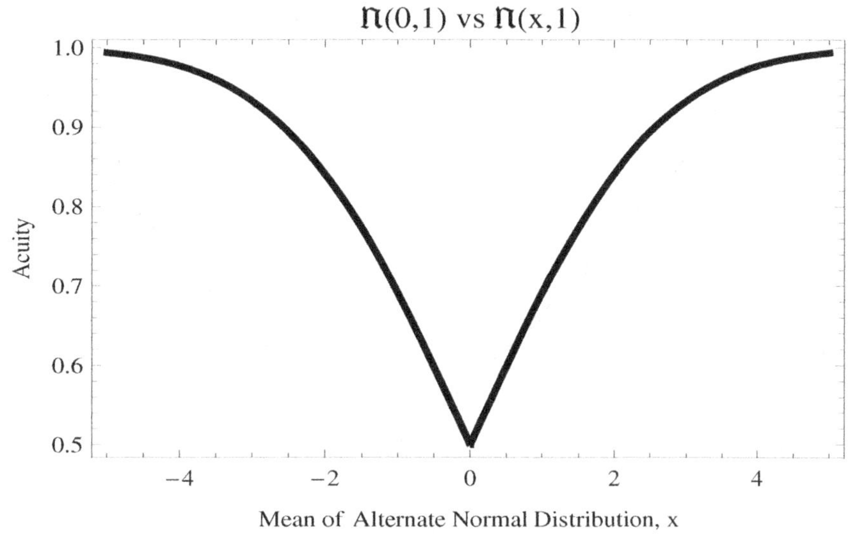

FIGURE 13.11. Acuity for $\mathfrak{N}(0, 1)$ vs $\mathfrak{N}(x, 1)$

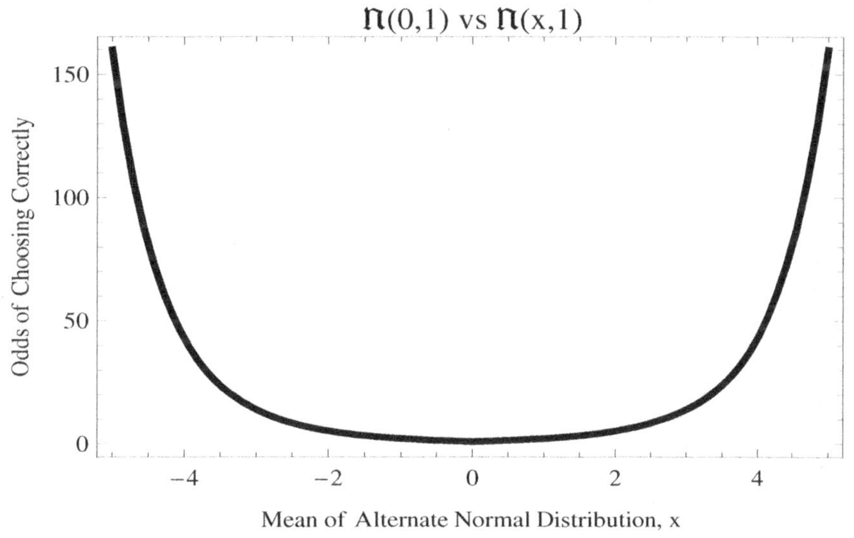

FIGURE 13.12. Odds for $\mathfrak{N}(0, 1)$ vs $\mathfrak{N}(x, 1)$

Figure 13.13 shows that the *logarithm* of the odds increase more or less linearly as x moves away from 0 at about $\frac{6}{5}$ of the difference in means. This is the reason that there are more often rigorous tests of hypotheses in the hard sciences than the soft.

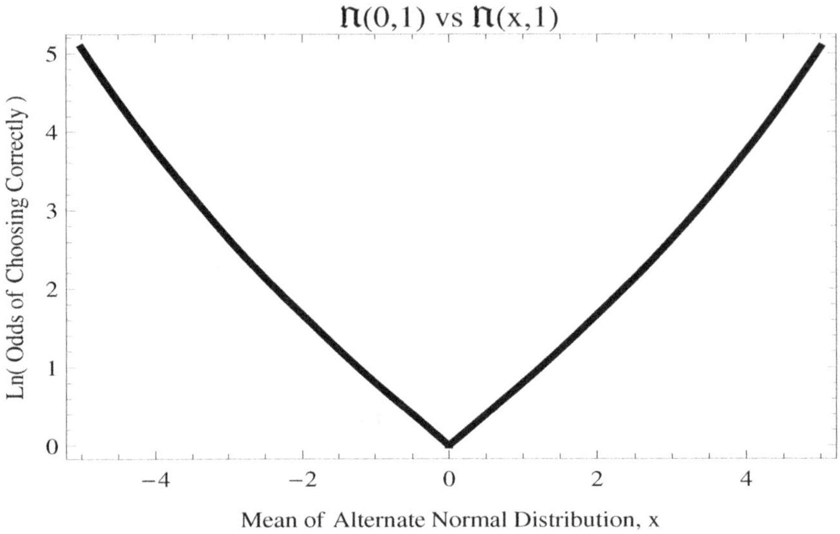

FIGURE 13.13. Logarithm of Odds for $\mathfrak{N}(0,1)$ vs $\mathfrak{N}(x,1)$

Figure 13.14 shows how the odds of choosing correctly between the two alternative hypotheses in the hard sciences sweeps upward and away from what can be accomplished in the soft sciences using a t-test. It all has to do with the form hypotheses! In the hard sciences, the experimenter has something specific in mind. In the soft sciences, the experimenter has little more than an idea that the result will be on one side or the other of zero. Because there are specific hypotheses in mind and usually the error term is a function of the instrumentation in the hard sciences, the odds of correctly selecting the true hypothesis is over 26 times as high in the hard sciences versus the soft sciences at the five standard deviation level.

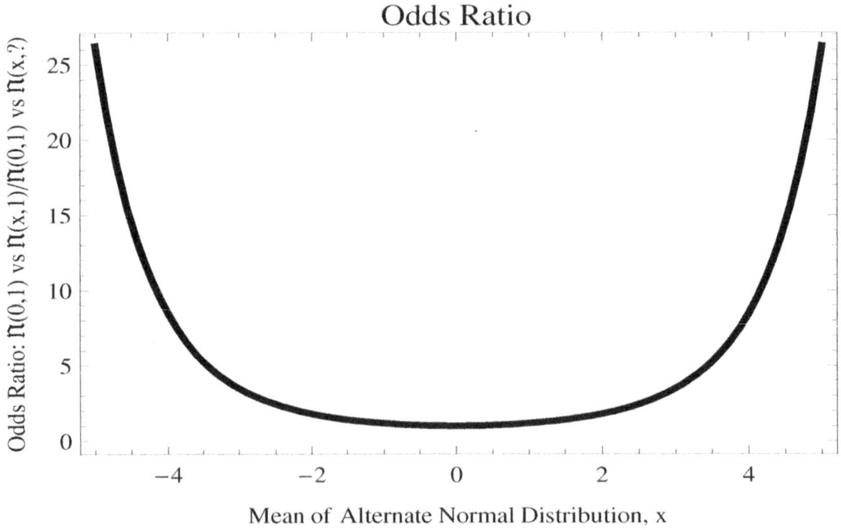

FIGURE 13.14. Odds Ratio for $\mathfrak{N}(0,1)$ vs $\mathfrak{N}(x,1)$ and $\mathfrak{N}(0,1)$ vs $\mathfrak{N}(x,?)$

Vague hypotheses are hard to validate. Significance testing has produced an inflated sense of certainty that is not justified and far below what is acceptable in the hard sciences. The quantitative differences are so great that the soft sciences are not science at all.

The core of the hard sciences could fit in a small booklet. There is no core in the soft sciences, just a mass of unsubstantiated hypotheses that could fill a library. There is a mass of published articles showing purported truths about behavior, diet, the causes of disease, *etc.* Replication is seldom attempted and when it is, it too often fails.[3]

In the next chapter, more difficult decisions are addressed—comparing the results of two complex models.

[3]See Makel, *et al.* [**32**], Nosek and Lakens [**38**], and the rest of Social Psychology, Volume 45, No 3, 2014.

CHAPTER 14

Goodness of Fit

Fitting Models

In this chapter, we will discover a way to manufacture two normal distributions and an optimum decision rule from the R^2 of a the estimation of a regression of a dependent variable on a number of independent variables. The normal distributions the can be derived from R^2 are, in turn, derived from the error of the regression equation and the standard deviation of the dependent variable from its mean.

We will use that result directly in Chapter 21 to examine the core findings of the book, *The Bell Curve.*

When we know where multiple distributions lie in relation to the same distribution, we know where they lie in relation to each other. This is the case when there are multiple regressions of the same dependent variable.

An example will be found later in Chapter 22 where we compare Laffer and Ranson's equations with policy variables and seasonal dummy variables to equations with *just* seasonal variables.

Along the way, we will see how sample size and R^2 interact with acuity in relationship the default null hypothesis.

For the default null hypothesis, the natural candidate is the hypothesis is that in which the independent variables have zero effect—that the target variable simply varies around a mean.

$$\mathfrak{h}_1: \quad y_i = \alpha + \beta_1 x_{1_i} + \beta_2 x_{2_i} + \cdots + \beta_k x_{k_i} + \varepsilon_i$$

$$\mu_y = \frac{\sum_{i=1}^n y_i}{n}$$

$$\sigma_y^2 = \frac{\sum_{i=1}^n (y_i - \mu_y)^2}{n-1}$$

$$\sigma_\varepsilon^2 = \frac{\sum_{i=1}^n \varepsilon_i^2}{n-k-1}$$

$$\varepsilon = \mathfrak{N}(0, \sigma_\varepsilon)$$

$$\blacktriangleright \rho = \frac{\sigma_\varepsilon^2}{\sigma_y^2} \blacktriangleleft$$

$$\mathfrak{h}_2: \quad \mu_y = \frac{\sum_{i=1}^n y_i}{n}$$

$$\sigma_y^2 = \frac{\sum_{i=1}^n (y_i - \mu_y)^2}{n-1}$$

$$y = \mathfrak{N}(\mu_y, \sigma_y)$$

One wants to put these two distributions side by side. There are two ways to look at this situation. The formula for ρ shows the way:

Case A: $\sum_{i=1}^n \frac{(y_i-\mu_y)^2}{\sigma_y^2}$ follows χ^2_{n-1} whoose mean is $= n-1$ and $\sum_{i=1}^n \frac{\varepsilon_i^2}{\sigma_\varepsilon^2}$ follows χ^2_{n-k-1} whose mean is $= (n-1)\rho$.

Case B: $\sum_{i=1}^n \frac{\varepsilon_i^2}{\sigma_\varepsilon^2}$ follows χ^2_{n-k-1} whose mean is $= n-k-1$ and $\sum_{i=1}^n \frac{(y_i-\mu_y)^2}{\sigma_y^2}$ follows χ^2_{n-1} whose mean is $= \frac{(n-k-1)}{\rho}$.

In either case, we have two χ^2 distributions and know the distance separating them, which is the same no matter which end of the telescope one is looking through.

$\mathfrak{h}_1$ and $\mathfrak{h}_2$ can be derived from R^2, $\rho = 1 - \overline{R}^2 = \frac{(n-1)(1-R^2)}{n-k-1}$.

To simplify matters, we assume that the sample size is large. Relying on the fact that $\chi^2_n \to \mathfrak{N}(n, \sqrt{2n})$ as $n \to \infty$, we transform a value of R^2 into two normal distributions. First taking note of the necessity of adjusting R^2 for the degrees of freedom lost in estimating both the model that is source of the R^2 and the "straw man" alternative. The adjusted R^2 is:

$$\overline{R}^2 = 1 - \frac{(n-1)(1-R^2)}{n-k-1} = 1 - \rho$$

In Cases A and B:

$$\chi^2_{n-1} \approx \mathfrak{N}(n-1, \sqrt{2(n-1)}) \text{ for large } n$$
$$\chi^2_{n-k-1} \approx \mathfrak{N}(n-k-1, \sqrt{2(n-k-1)}) \text{ for large } n$$

where n is the number of data points and k is the number of parameters estimated in the model, not including the mean of variable being modeled. We choose to, without loss of generality, to scale the two distributions consistent with Case A.

Because the sample size is large enough, as an approximation in the following diagrams we use these two normal distributions:

$$\mathfrak{N}\left((n-1)(1-\overline{R}^2), \sqrt{2(n-k-1)}\right) \text{ vs. } \mathfrak{N}\left(n-1, \sqrt{2(n-1)}\right)$$

to formulate optimum decision rule and calculate its acuity.

Varying R^2

Figures 14.1 to 14.4 show how, even with small sample size, a strong result can be obtained. Unfortunately such high R^2 seldom, if ever, occurs in the soft sciences. It is hard to fell comfortable with these diagrams as offering a choice between hypotheses because the result is preordained by the process of fitting the model. Imagine, however, that this was a replication and the value of ρ was determined by a previous study. In that case, the choice would be open between the model, given new data and old coefficients, producing errors consistent with the old fit or nearer to or perhaps greater than the new σ_y.

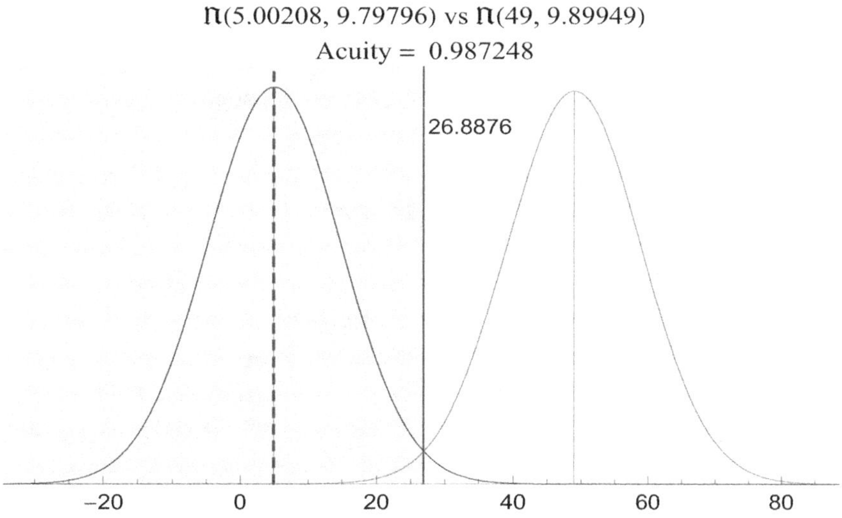

FIGURE 14.1. $n = 50$ and R^2 – .90

At this point, the Acuity produces odds consistent with the standards of the hard sciences.

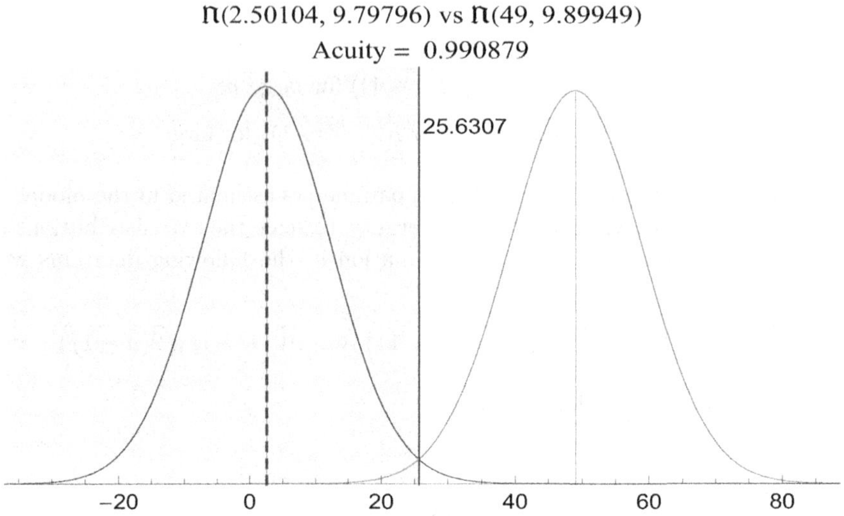

FIGURE 14.2. $n = 50$ and $R^2 = .95$

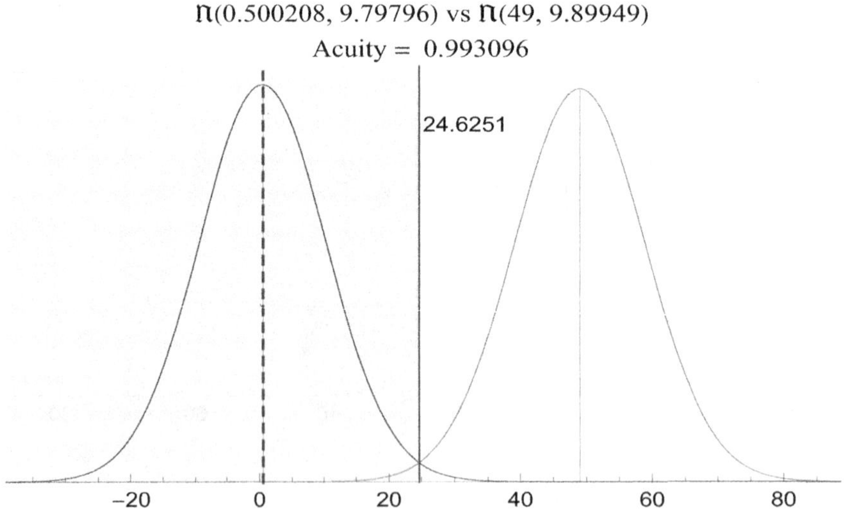

FIGURE 14.3. $n = 50$ and $R^2 = .99$

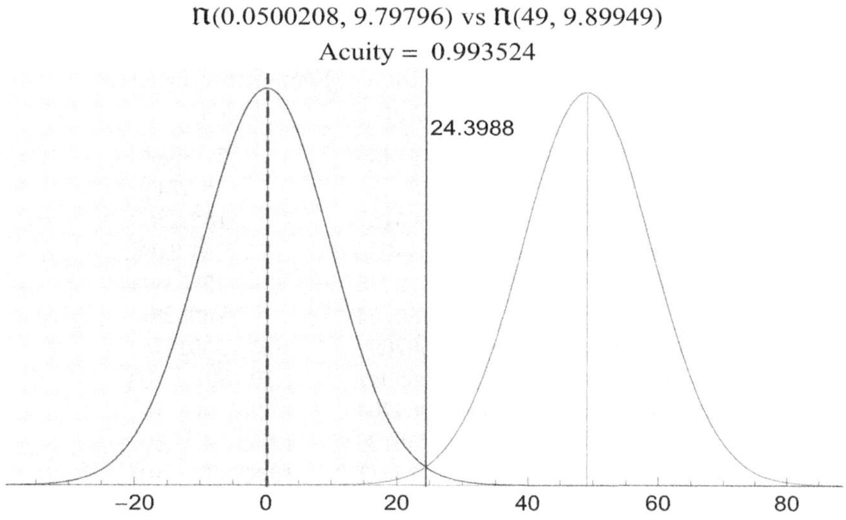

FIGURE 14.4. $n = 50$ and $R^2 = .999$

Table 14.1 on the following page shows how the inflated the significance of significance levels distorts judgment. The researcher in the soft sciences would be led to believe that, with a sample size of 50, he would achieve a certainty like that standard in the hard sciences with R^2 of somewhere between .10 and .15, when a level above 0.999999 is required. With a sample of 50, being of a par with the hard sciences is almost out of reach.

One Variable Regression; sample size = 50			
	Optimum Decision		Significance Level
R^2	Acuity	Odds	Odds
0.050000	0.5299512	1.127439	7.436886
0.100000	0.5800886	1.381455	38.57436
0.150000	0.6289573	1.695107	182.2280
0.200000	0.6758286	2.084788	884.2166
0.250000	0.7200663	2.572275	4585.894
0.300000	0.7611515	3.186754	26032.60
0.350000	0.7986993	3.967692	165177.8
0.400000	0.8324663	4.968947	1.198246×10^{6}
0.450000	0.8623481	6.264704	1.020706×10^{7}
0.500000	0.8883695	7.958128	1.055486×10^{8}
0.550000	0.9106673	10.19411	1.383157×10^{9}
0.600000	0.9294691	13.17818	2.433296×10^{10}
0.650000	0.9450698	17.20494	6.227710×10^{11}
0.700000	0.9578078	22.70105	2.607537×10^{13}
0.750000	0.9680421	30.29118	2.141723×10^{15}
0.800000	0.9761336	40.89993	4.676421×10^{17}
0.850000	0.9824288	55.91115	4.797132×10^{20}
0.900000	0.9872481	77.41976	8.298959×10^{24}
0.950000	0.9908787	108.6339	1.428828×10^{32}
0.975000	0.9923278	129.3400	2.427207×10^{39}
0.990000	0.9930963	143.8503	8.686556×10^{48}
0.995000	0.9933370	149.0816	1.460890×10^{56}
0.999000	0.9935241	153.4186	8.724364×10^{72}
0.999900	0.9935656	154.4136	8.728136×10^{96}
0.999990	0.9935697	154.5135	8.728513×10^{120}
0.999999	0.9935701	154.5235	8.728551×10^{144}

TABLE 14.1. How to deceive yourself with R^2.

Varying n

This section examines the effect of increasing sample size on a model with $R^2 = .05$. Figures 14.5 through 14.12 show that extremely large sample sizes are required to provide convincing evidence for a weak effect.

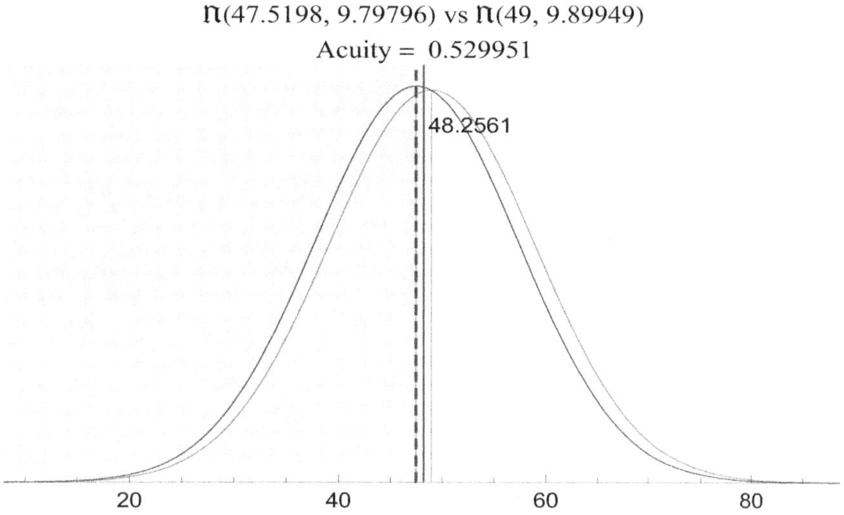

FIGURE 14.5. $n = 50$ and $R^2 = .05$

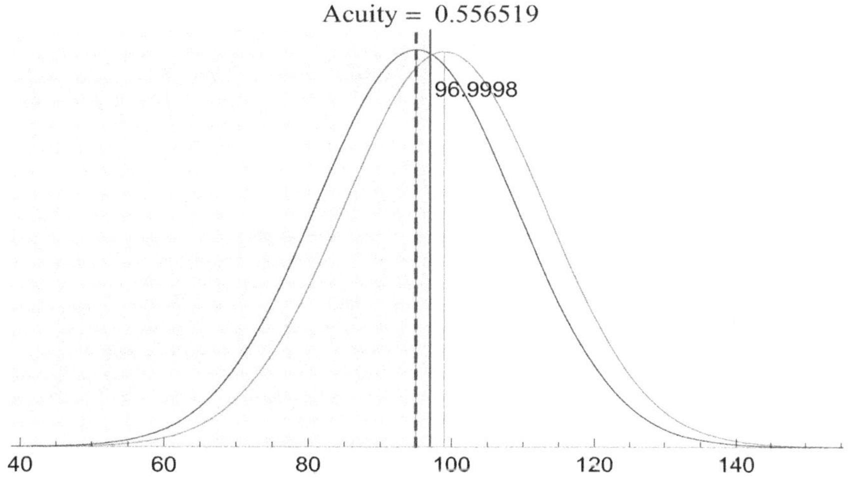

FIGURE 14.6. $n = 100$ and $R^2 = .05$

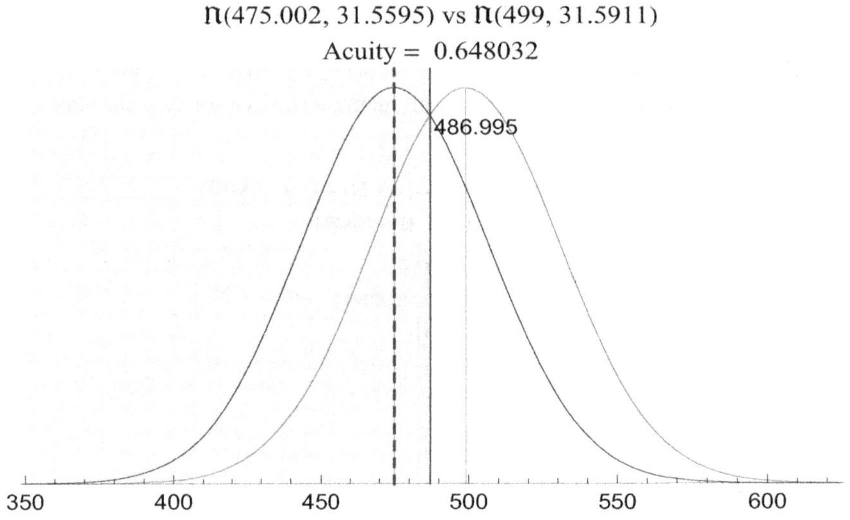

FIGURE 14.7. $n = 500$ and $R^2 = .05$

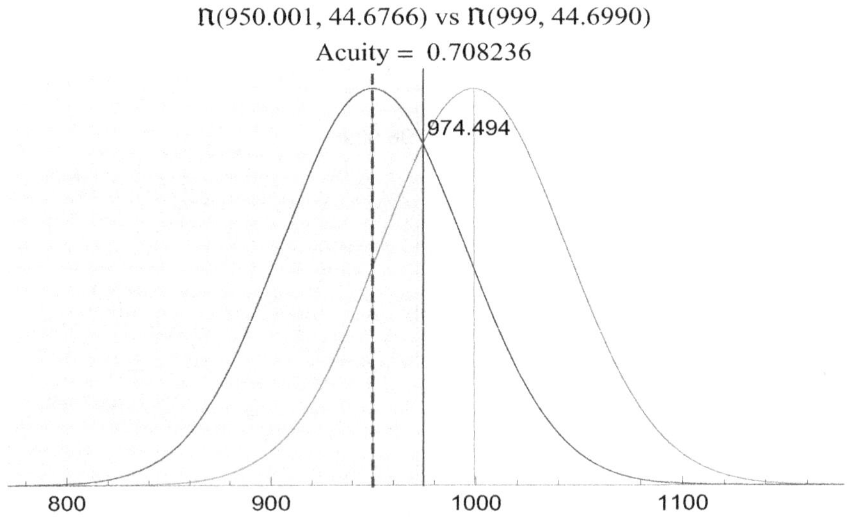

FIGURE 14.8. $n = 1{,}000$ and $R^2 = .05$

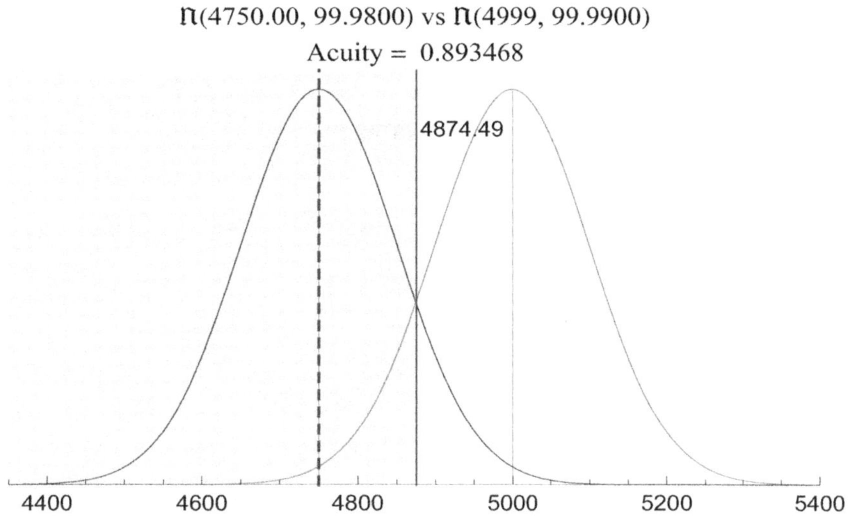

FIGURE 14.9. $n = 5{,}000$ and $R^2 = .05$

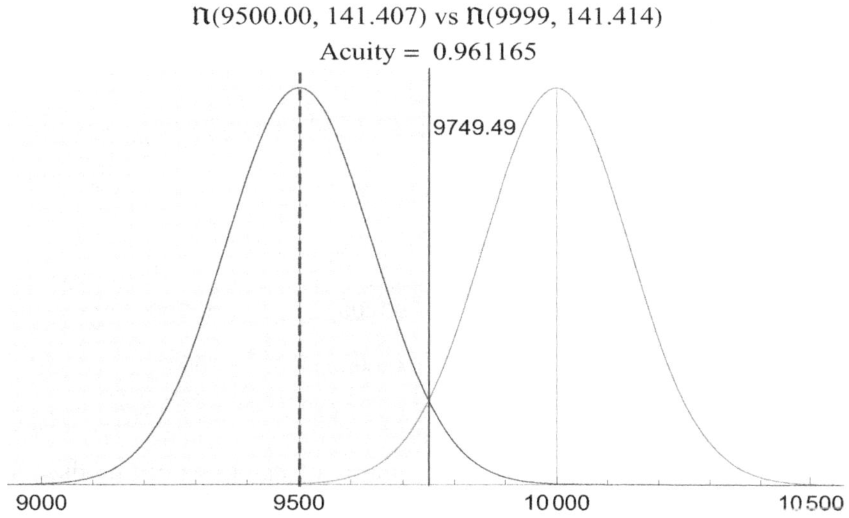

FIGURE 14.10. $n = 10{,}000$ and $R^2 = .05$

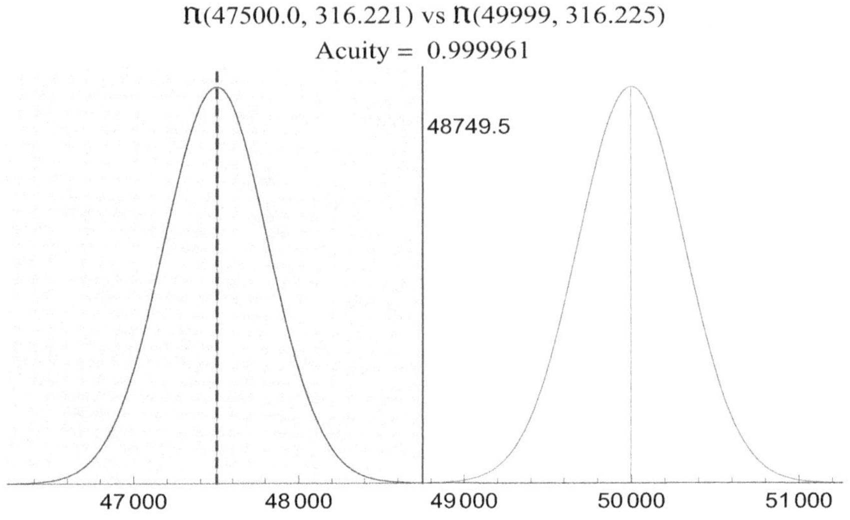

FIGURE 14.11. n = 50,000 and R^2 = .05

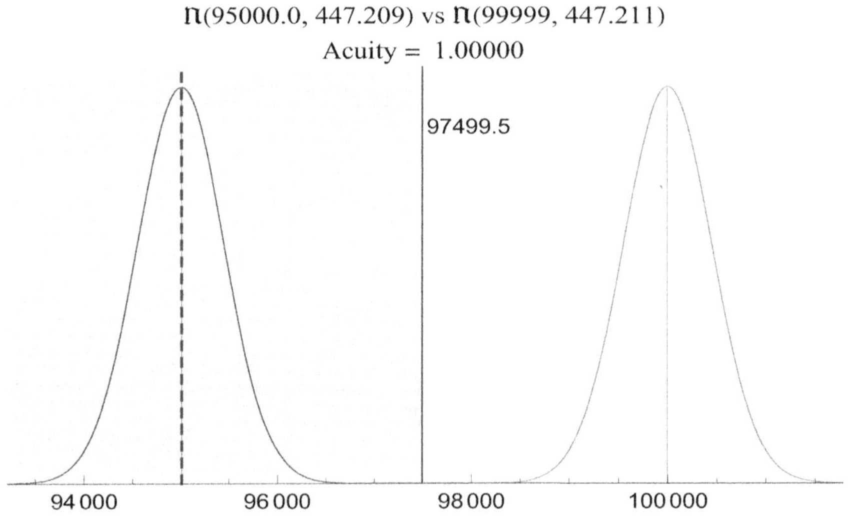

FIGURE 14.12. n = 100,000 and R^2 = .05

Table 14.2 on the next page shows to what lengths one must in sample size to produce a decisive answer for an hypothesis of modest R^2.

A practitioner of the soft sciences, if he misinterpreted the meaning of significance level, might think he was reaching a level of certainty equal to the hard sciences with a sample size somewhere between 100 and 500. In reality, to achieve to goal a sample size between 10,0000 and 50,000 is required—again where the acuity equals 0.99379.

Lackluster hypotheses and small samples mean results that should not be believed.

Effect of Varying Sample Size with $R^2 = .05$			
	Optimum Decision		**Significance Level**
n	**Acuity**	**Odds**	**Odds**
50	0.5299512	1.127439	7.436886
100	0.5565189	1.254888	38.48090
500	0.6480324	1.841170	2.282306×10^{6}
1000	0.7082355	2.427422	1.177695×10^{12}
5000	0.8934683	8.386875	9.274919×10^{56}
10000	0.9611646	24.74970	6.427333×10^{112}
50000	0.9999611	25732.52	4.839268×10^{558}
100000	1.000000	8.762030×10^{7}	$5.560109 \times 10^{1115}$

TABLE 14.2. How to deceive yourself with sample size.

Figure 14.13 shows the interaction between sample size and R^2 on the odds of choosing correctly between the model and just fluctuating around a mean. An examination of the graph show that log(Odds from Acuity) are linear with increases in sample size and approximately exponential with respect to R^2.

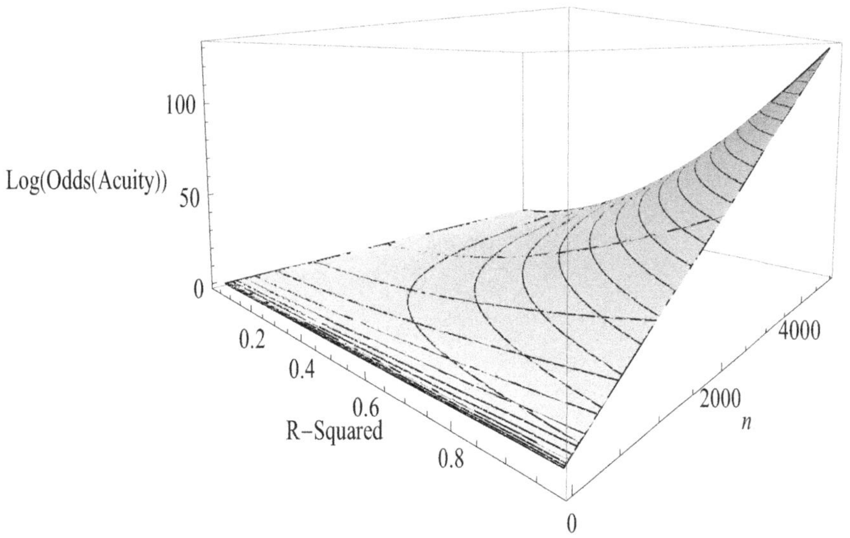

FIGURE 14.13. Log_{10}(Odds from Acuity) vs. n and R^2

It is clear that only extremely high R^2 can provide evidence for the probability that a regression is true versus the straw-man alternative with modest sample sizes. Strong effects are seldom seen in the soft sciences and sufficiently large sample sizes are seldom thought necessary because statistical significance overstates the certainty of the results.

Part III

Compound Hypotheses & Replication

CHAPTER 15

Introduction to Part III

The most basic statistical question is analyzing the difference between two populations. It naturally arises in medical, epidemiological, psychological, and sociological fields. It arises wherever treatments, interventions, causes of disease, or psychological traits are hypothesized to have an effect. It has been incorrectly analyzed for a hundred years.

The correct procedure is to treat the question as a compound hypothesis—the proposition that several independent hypotheses are all true.

Chapter 16 will reanalyze the first time the incorrect procedure was used, the classic paper by Ronald Fisher[**15**] that started the snowball rolling down into the wrong valley.

The other instance where a compound hypothesis arises is in the replication of an experiment. The same chapter will describe the possible outcomes if Fisher had replicated his experiment.

Chapter 17 will describe a second instance of replication, the replication of a regression analysis or analysis of variance—whatever can be summarized by the percentage of explained variation.

Other Distributions Become Normal Distributions as $n \to \infty$.

In the following section we will need justification for using normal distribuions as proxies for other distributions. I reference *Statistical Distributiibs*, Fourth Edition by Catherine Forbes, Merran Evans, Nicholas Hastings, and Brian Peacock, Chapter 33, Page 146

CHAPTER 16

Fisher's Experiment

In 1925 Sir Ronald Fisher published a paper [**15**] filling some gaps in "Student's" derivation of the t-distribution and clarifying its usage. As an exemplar, he did the calculations for an experiment done at the Rothamsed Experimental Station in 1922. Barley plants were divided into two groups. All were exposed to "high tension" electrical discharge, but some were shielded by an earthed wire cage and some were not.[1].

The number of shoots were counted. The mean number of shoots of shielded plants was 23.333, m_1, that of the unshielded was 17.625, m_2. Fisher then applied a two-tailed t-test to the difference in of the means, finessing the difference in standard deviations the way people have done ever since, and concluded "that this value will be exceeded by chance about 41 times in 10,000." $p = 0.0041$.

Redoing the math, his statement is equivalent the statement:

> If the average difference between the treatments is either 5.7083 with a standard deviation that average $= 1.8688$, $\mathfrak{h}_1$, or the average difference between the treatments is zero, $\mathfrak{h}_2$; there is a .80624 probability that the $\mathfrak{h}_1$ is true.

Now there is an enormous difference between

$$p = 0.0041 \text{ and } \mathrm{P}(\text{“}\mathfrak{h}_1 \text{is true”} \mid (m_1 \;\&\; m_2) \;\&\; (\mathfrak{h}_1 \vee \mathfrak{h}_2)) = .80624$$

The results are shown in Figure 16.1 and Table 16.1. One should not be surprised at this. The answer to a question depends on the question asked. Fisher's answer does not address the question of which alternative is true; the probability of $\mathfrak{h}_1$ being true is only surprising because one has spent so much time looking at the results of significance tests.

[1]For those wishing to read the original paper, be aware that "tillers" means "shoots" and S is used where we would now use Σ

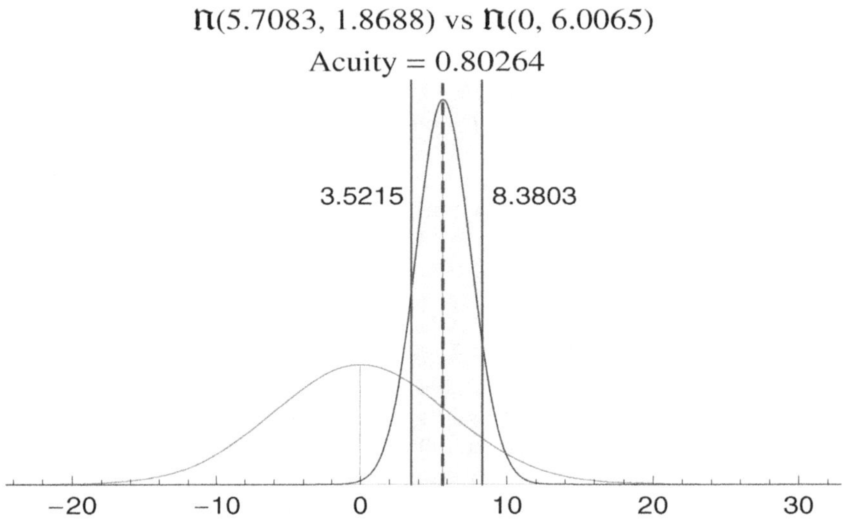

FIGURE 16.1. Acuity of Fisher's Experiment using the assumptions of the t-test

	Fisher's Experiment using the assumptions of the t-test Shielding vs. No Shielding	
	3.5215 < m > 8.3803	**m < 3.5215 or m > 8.3803**
$\mathfrak{h}_1$: Shielding makes a difference	.80264	.19736
$\mathfrak{h}_2$: Shielding makes no difference	.19736	.80264
	say: "$\mathfrak{h}_1$ is true"	say: "$\mathfrak{h}_2$ is true"
	The Decision: Acuity = .80264	

TABLE 16.1. Fisher's Experiment using the assumptions of the t-test

This is not he end of the story. The t-test ignores some of the information available—the individual means and standard deviations of the shielded and unshielded samples. The shielded sample was

$$\mathfrak{N}(m_{sh}, s_{sh}) = \mathfrak{N}(23.333, 1.6956)$$

and unshielded was

$$\mathfrak{N}(m_{un}, s_{un}) = \mathfrak{N}(17.625, 0.87918)$$

Let us hypothesize that

$$\mathfrak{h}_{Shielded} = \mathfrak{N}(\mu_{sh}, \sigma_{sh}) = \mathfrak{N}(m_{sh}, s_{sh})$$

and

$$\mathfrak{h}_{Unshielded} = \mathfrak{N}(\mu_{un}, \sigma_{un}) = \mathfrak{N}(m_{un}, s_{un})$$

What we want to show to what extent that $\mathfrak{h}_{Shielded}$ and $\mathfrak{h}_{Unhielded}$ are *both* true based on the shielded and unshielded samples. The first step is to calculate the decision rule, $\boldsymbol{d}$, that maximizes the the $\boldsymbol{\mathcal{A}_d}$ of $\mathfrak{N}(23.333, 1.6956)$ vs. $\mathfrak{N}(17.625, 0.87918)$. That result is shown in Figure 16.2, $\boldsymbol{\mathcal{A}_d} = .98545$.

Since we have to use the decision rule twice, once for the shielded sample and once for the unshielded sample, and since the two samples are independent of each other, the combined acuity is the product of the individual acuities.

$$\boldsymbol{\mathcal{A}}_{(Shielded\ \&\ Unshielded)} = \boldsymbol{\mathcal{A}}_{Shielded} \times \boldsymbol{\mathcal{A}}_{Unshielded} = (.98545)^2 = .971102$$

The distribution of probability in the counterfactual combinations is shown in Tables 16.3 and 16.4

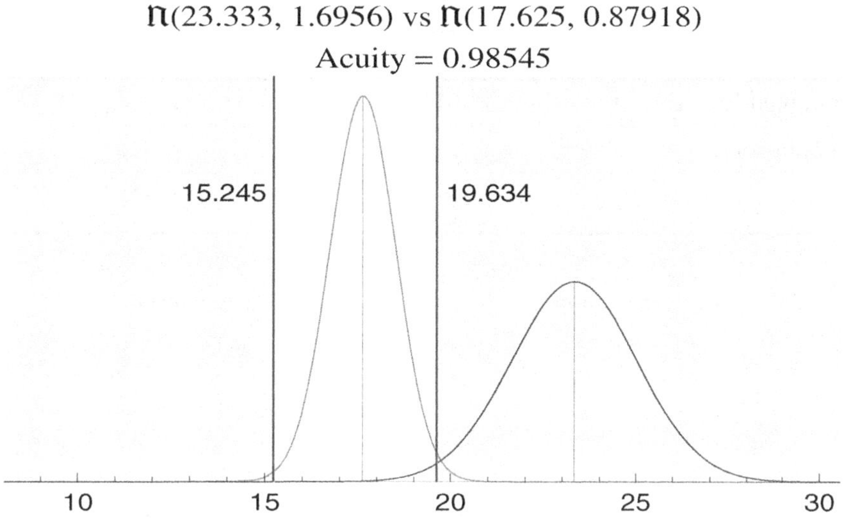

FIGURE 16.2. $\mathfrak{h}_{Shielded}$ vs. $\mathfrak{h}_{Unshielded}$

Suppose that Fisher had repeated the experiment and although the results were not identical, the same decision rule, $\boldsymbol{d}$ mapped the new results into the right places:

$$\boldsymbol{d}\colon m_{sh} \to \mathfrak{h}_{Shielded} \text{ and } \boldsymbol{d}\colon m_{un} \to \mathfrak{h}_{Unshielded}$$

	$\boldsymbol{d}\colon m_{sh} \to \mathfrak{h}_{Shielded}$	$\boldsymbol{d}\colon m_{sh} \to \mathfrak{h}_{Unshielded}$
$\boldsymbol{d}\colon m_{un} \to \mathfrak{h}_{Unshielded}$	$\boldsymbol{\mathcal{A}}_d^2$	$\boldsymbol{\mathcal{A}_d} \times (1 - \boldsymbol{\mathcal{A}_d})$
$\boldsymbol{d}\colon m_{un} \to \mathfrak{h}_{Shielded}$	$\boldsymbol{\mathcal{A}_d} \times (1 - \boldsymbol{\mathcal{A}_d})$	$(1 - \boldsymbol{\mathcal{A}_d})^2$

TABLE 16.2. Probability of ($\mathfrak{h}_{Shielded}$ & $\mathfrak{h}_{Unshielded}$)

The results are shown in Tables 16.4 and 16.5.

	$\boldsymbol{d}\colon m_{sh} \to \mathfrak{h}_{Shielded}$	$\boldsymbol{d}\colon m_{sh} \to \mathfrak{h}_{Unshielded}$
$\boldsymbol{d}\colon m_{un} \to \mathfrak{h}_{Unshielded}$	0.971102	0.0143431
$\boldsymbol{d}\colon m_{un} \to \mathfrak{h}_{Shielded}$	0.0143431	0.000211847

TABLE 16.3. Numerical Values

	$\mathfrak{h}_{Shielded}$ & $\mathfrak{h}_{Unshielded}$	$\neg(\mathfrak{h}_{Shielded}$ & $\mathfrak{h}_{Unshielded})$
Original Experiment	$\boldsymbol{\mathcal{A}}_d^2$	$1 - \boldsymbol{\mathcal{A}}_d^2$
Replication	$\boldsymbol{\mathcal{A}}_d^2$	$1 - \boldsymbol{\mathcal{A}}_d^2$
Combined Probability	$\frac{\boldsymbol{\mathcal{A}}_d^4}{\boldsymbol{\mathcal{A}}_d^4+(1-\boldsymbol{\mathcal{A}}_d^2)^2}$	$\frac{(1-\boldsymbol{\mathcal{A}}_d^2)^2}{\boldsymbol{\mathcal{A}}_d^4+(1-\boldsymbol{\mathcal{A}}_d^2)^2}$

TABLE 16.4. Formulas for the replication of ($\mathfrak{h}_{Shielded}$ & $\mathfrak{h}_{Unshielded}$)

	$\mathfrak{h}_{Shielded}$ & $\mathfrak{h}_{Unshielded}$	$\neg(\mathfrak{h}_{Shielded}$ & $\mathfrak{h}_{Unshielded})$
Original Experiment	0.971102	0.028898
Replication	0.971102	0.028898
Combined Probability	0.999115	0.000885

TABLE 16.5. Numerical Values

Suppose, however, the results of the replication were different. Suppose that:

$$\boldsymbol{d}\colon m_{sh} \to \mathfrak{h}_{Unshielded} \text{ and } \boldsymbol{d}\colon m_{un} \to \mathfrak{h}_{Unshielded}$$

or

$$\boldsymbol{d}\colon m_{sh} \to \mathfrak{h}_{Shielded} \text{ and } \boldsymbol{d}\colon m_{un} \to \mathfrak{h}_{Shielded}$$

The results are shown in Tables 16.6 & 16.7.

	$\mathfrak{h}_{Shielded}$ & $\mathfrak{h}_{Unshielded}$	$\neg(\mathfrak{h}_{Shielded}$ & $\mathfrak{h}_{Unshielded})$
Original Experiment	$\boldsymbol{\mathcal{A}}_d^2$	$1 - \boldsymbol{\mathcal{A}}_d^2$
Replication	$\boldsymbol{\mathcal{A}}_d \times (1 - \boldsymbol{\mathcal{A}}_d)$	$1 - (\boldsymbol{\mathcal{A}}_d \times (1 - \boldsymbol{\mathcal{A}}_d))$
Combined Probability	$\frac{\boldsymbol{\mathcal{A}}_d^3 - \boldsymbol{\mathcal{A}}_d^4}{1 - \boldsymbol{\mathcal{A}}_d + 2\boldsymbol{\mathcal{A}}_d^3 - 2\boldsymbol{\mathcal{A}}_d^4}$	$\frac{1 - \boldsymbol{\mathcal{A}}_d + \boldsymbol{\mathcal{A}}_d^3 - \boldsymbol{\mathcal{A}}_d^4}{1 - \boldsymbol{\mathcal{A}}_d + 2\boldsymbol{\mathcal{A}}_d^3 - 2\boldsymbol{\mathcal{A}}_d^4}$

TABLE 16.6. Formulas for the replication of ($\mathfrak{h}_{Shielded}$ & $\mathfrak{h}_{Unshielded}$)

	$\mathfrak{h}_{Shielded}$ & $\mathfrak{h}_{Unshielded}$	$\neg(\mathfrak{h}_{Shielded}$ & $\mathfrak{h}_{Unshielded})$
Original Experiment	0.971102	0.028898
Replication	0.014343	0.985657
Combined Probability	0.328411	0.671589

TABLE 16.7. Numerical Values

Suppose that the worst happened—the second was *shielded* like the original *unshielded* and the second *shielded* was like the original *unshielded.*

$$\boldsymbol{d}\colon m_{sh} \rightarrow \mathfrak{h}_{Unshielded} \text{ and } \boldsymbol{d}\colon m_{un} \rightarrow \mathfrak{h}_{Shielded}$$

	$\mathfrak{h}_{Shielded}$ & $\mathfrak{h}_{Unshielded}$	$\neg(\mathfrak{h}_{Shielded}$ & $\mathfrak{h}_{Unshielded})$
Original Experiment	$\boldsymbol{\mathcal{A}}_d^2$	$1 - \boldsymbol{\mathcal{A}}_d^2$
Replication	$(1 - \boldsymbol{\mathcal{A}}_d)^2$	$1 - (1 - \boldsymbol{\mathcal{A}}_d)^2$
Combined Probability	$\frac{\boldsymbol{\mathcal{A}}_d^2 - 2\boldsymbol{\mathcal{A}}_d^3 + \boldsymbol{\mathcal{A}}_d^4}{2\boldsymbol{\mathcal{A}}_d - 4\boldsymbol{\mathcal{A}}_d^3 + 2\boldsymbol{\mathcal{A}}_d^4}$	$\frac{2\boldsymbol{\mathcal{A}}_d - \boldsymbol{\mathcal{A}}_d^2 - 2\boldsymbol{\mathcal{A}}_d^3 + \boldsymbol{\mathcal{A}}_d^4}{2\boldsymbol{\mathcal{A}}_d - 4\boldsymbol{\mathcal{A}}_d^3 + 2\boldsymbol{\mathcal{A}}_d^4}$

TABLE 16.8. Formulas for the replication of ($\mathfrak{h}_{Shielded}$ & $\mathfrak{h}_{Unshielded}$)

	$\mathfrak{h}_{Shielded}$ & $\mathfrak{h}_{Unshielded}$	$\neg(\mathfrak{h}_{Shielded}$ & $\mathfrak{h}_{Unshielded})$
Original Experiment	0.971102	0.028898
Replication	0.000212	0.999788
Combined Probability	0.007070	0.992930

TABLE 16.9. Numerical Values

CHAPTER 17

Replication Tests of Regression

Suppose that $\widehat{\mathfrak{h}}_1$ is a model of y that is being replicated using a new dataset of size, m. The procedure is to use the coefficients from the old estimation applied to the new data set, hoping that as much of the variation in the new data will be explained as occurred with the previous estimation.

Let $\hat{x}_{k_i}$ is the ith observation of the kth independent variable and $\hat{y}_i$ be the ith observation of dependent variable.

$$\widehat{\mathfrak{h}}_1: \quad \hat{y}_i = \hat{\alpha} + \hat{\beta}_1\hat{x}_{1_i} + \hat{\beta}_2\hat{x}_{2_i} + \cdots + \hat{\beta}_k\hat{x}_{k_i} + \hat{\varepsilon}_i$$

$$\mu_{\hat{y}} = \frac{\sum_{i=1}^{n}\hat{y}_i}{n}$$

$$\sigma_{\hat{y}}^2 = \frac{\sum_{i=1}^{n}(\hat{y}_i - \mu_{\hat{y}})^2}{n-k-1}$$

$$\sigma_{\hat{\varepsilon}}^2 = \frac{\sum_{i=1}^{n}\hat{\varepsilon}_i^2}{n}$$

$$\hat{\varepsilon} = \mathfrak{N}(0, \sigma_{\hat{\varepsilon}})$$

$$\boxed{\hat{\rho} = \frac{\sigma_{\hat{\varepsilon}}^2}{\sigma_{\hat{y}}^2}}$$

Let $\{y_i, x_{1_i}, \ldots, x_{k_i}\}$ be a *new* set of data with m observations and $\hat{\beta}_1, \ldots, \hat{\beta}_k$ are the unchanged coefficients from the previous regression.

$$\mathfrak{h}_1: \quad y_i = \alpha + \hat{\beta}_1 x_{1_i} + \hat{\beta}_2 x_{2_i} + \cdots + \hat{\beta}_k x_{k_i} + \varepsilon_i$$

$$\alpha = \frac{\sum_{i=1}^{m} y_i - (\hat{\beta}_1 x_{1_i} + \hat{\beta}_2 x_{2_i} + \cdots + \hat{\beta}_k x_{k_i})}{m}$$

$$\mu_y = \frac{\sum_{i=1}^{m} y_i}{m}$$

$$\sigma_y^2 = \frac{\sum_{i=1}^{m}(y_i - \mu_y)^2}{m-1}$$

$$\sigma_\varepsilon^2 = \frac{\sum_{i=1}^{m}\varepsilon_i^2}{m}$$

$$\varepsilon = \mathfrak{N}(0, \sigma_\varepsilon)$$

$$\boxed{\sigma_\varepsilon^2 \stackrel{?}{=} \hat{\rho}\,\sigma_y^2}$$

The alternative hypothesis is:

$$\mathfrak{h}_2: \quad \mu_{\hat{y}} = \frac{\sum_{i=1}^{n} \hat{y}_i}{n}$$
$$\sigma_y = \frac{\sum_{i=1}^{m}(y_i - \mu_y)^2}{m-1}$$
$$y = \mathfrak{N}(\mu_y, \sigma_y)$$

$$\sum_{i=1}^{m} \frac{(y_i - \mu_y)^2}{\sigma_y^2} \text{ follows } \chi^2_{m-1} \approx \mathfrak{N}(m-1, \sqrt{2(m-1)}) \text{ for large } m.$$

$$\sum_{i=1}^{m} \frac{\varepsilon_i^2}{\sigma_\varepsilon^2} \text{ follows } \chi^2_{m} \approx \mathfrak{N}(m, \sqrt{2(m)}) \text{ for large } m.$$

The test statistic is $\frac{\sum_{i=1}^{m} \varepsilon_i^2}{\sigma_y}$ and it should be approximately equal to $\hat{\rho}(m-1)$. If the compound probabilities are calculated, they are calculated just like they were in hypothetical replication of Fisher's experiment, Chapter 16.

I can almost hear the squeals of distress coming from the soft science departments of numerous universities. The tacit belief is that one cannot expect a model to fit new data with old parameters. I think it's generally believed that the best one can hope for is to fit the model to new data and have the parameters be somewhat similar.

There are three reasons one should not follow that path:

(1) Fitting new data with old parameters is what it means for a model to be true.
(2) The target is quite wide. The old parameters applied to the new data need explain only a little over half the variance explained in the original estimation.
(3) There is much to be gained from replication. For instance, a regression with one independent variable and sample size 50 that produces at fit with $R^2 = .40$ may look impressive, but it is not. If the researcher mistakenly derived odds from the significance level, he might act as though the odds against the null hypothesis being true are over 1 million to one, the odds against being hit by lightening in a single year. In fact the odds against the null hypothesis are about those of not losing at Russian roulette. If the parameters passed the second test of equal size the odds would go up to 24.26 to one. If the parameters passed a third test, the odds would go up to 119.4 to one.

If the associations in the soft sciences are all as weak as commonly believed, the way for results to gain the epistemological *bona fides* of the hard sciences is successful replication.

Part IV

Case Studies

CHAPTER 18

Introduction to Part IV

Science is like cheese. There are three types: hard, semi-soft, and soft. There are usually distinguised by subject matter, physics and chemistry being the hardest and psychology, economics, epidemiology, and sociology being the softest. This is not the correct distinction. Any of the sciences has the potential to be as hard as chemistry or physics. In truth, the distinction should be based on the form of hypotheses proposed.

Imagine that Newton's laws of gravitation were something like this. "Objects attract each other." You wouldn't know whether the force was constant in time, or varied with the density of the objects or the color of the objects is the distance between them or their mass. You wouldn't know whether they followed a straight line, acted like a knuckleball or a curveball, or jump to little quantum leaps towards each other. You wouldn't know whether the force was strong or weak or whether the strength or weakness depended on the chemical composition of the objects.

How could you possibly test it? The answer is that there *is* no way to test it and achieve a result that would be valid.

Suppose Newton's laws of gravitation went something like this: "objects attract each other and the strength of the attraction depends on some function of the masses of the two objects and some function of the distance between them." You would have a little bit more information but not really enough to put the statement to a test.

Without more precise information, there is no way to test such an hypothesis!

If hypotheses are precise, evidence is powerful in deciding between them. If hypotheses are vague, evidence provides little basis for deciding between them. In the six examples that follow, we will travel from hard, through semi-soft, to quite soft indeed and finally turn the new methods back on the established statistical methods.

In Chapter 19, we examine the Eddington solar eclipse expedition to test Einstein's general theory of relativity. It was at the time, considered to have produced decisive evidence in favor of the theory. The new methods to put a number to that decisiveness.

Chapter 20 takes up a question from the realm of economics, to what extent one is justified in believing that a nearly 200 year history of stock market returns justifies a belief that, that record is indicative of a random walk with upward bias (after inflation) with mean and standard deviation equal to the sample data rather than no positive return at all.

The third example, in Chapter 21, examines the substance of the argument made in *The Bell Curve*. That work put forth the argument that heredity largely determines intelligence, as measured on standardized tests, and that measured intelligence largely determines success in life. This last link in the argument is put to the test.

The fourth example, Chapter 22, tests an econometric model of the United States economy proposed by Laffer and Ranson against no model at all.

The fifth example, Chapter 23, examines evidence for paranormal precognition by Bem. Should one be a believer in parapsychology?

The sixth example, Chapter 24, turns the new methods loose on the practice of statistics itself. The claims of significance testing are pitted against Meehl's conjecture that with large enough sample size, 100% of two-tailed significance tests and 50% of one-tailed tests will be seen to be significant at whatever level of significance.

CHAPTER 19

The Eddington Expedition

In 1916 Albert Einstein, based on his recently presented General Theory of Relativity [**12**], predicted that starlight would be bent by the gravitational field of the sun. In 1911, based on his Special Theory of Relativity, he had made a similar, but numerically different prediction [**11**]. In 1919, two British expeditions, organized by Frank Watson Dyson and Arthur Stanley Eddington, were dispatched with equipment to test his hypothesis by photographing the apparent position of stars near the sun on the stellar field during a total eclipse of the sun. In March of 1919, they set sail from England in order to be in place. A. Crommelin and C. Davidson went to Sobral, Brazil. Eddington and E. Cottingham went to Principe, an island off the coast of West Africa.

The Sobral group took two telescopes, an astrographic telescope and a backup 4-inch telescope, enabling them to take two sets of measurements at the time of the eclipse. The Principe group took a single instrument. Their task was to photograph stars near the edge of the sun, at the time of totality. Later these photographs were compared to photographs made of the section of the sky when the sun was elsewhere. If the sun's mass caused the starlight to bend toward the sun, the apparent position of the stars would shift away from the sun.

In the report to the Royal Astronomical Society [**8**], Dyson, Eddington, and C. Davidson considered three alternative hypotheses, to be tested against the data from the two eclipse expeditions. The first of these was, simply, that light traveled in a straight line, that there would be no deflection. This was in accord with Newtonian optics. The second possibility noted was that light had mass as indicated by Special Relativity, $m = \frac{E}{c^2}$, and that mass followed the Newtonian equations of gravitation. In the second case, the deflection was calculated to be $0.875''$ of an arc rather than zero, as in the first case. The third possibility was that in the light had mass was effected by gravitation, but in addition, time was slowed down by the sun's mass as called for by General Relativity. In this case, the deflection was calculated to be $1.75''$.

There were many difficulties collecting the data. Some of the data sets were smaller that desired because of cloud cover at totality. Some of the photographs were less clear than desired because of defects in the optics or the tracking mechanism. The best data was from the smaller telescope at Sobral.

In reading the published report, one cannot but be astounded how much the protocols for reporting experimental findings have changed in the intervening century. For one thing, the report contains vastly more detail than is common today. On cannot help but think the old procedure is the better.[1] In the instance, standards of reporting statistical results and knowledge of statistical procedures have changed greatly. Because of the level of detail, Earman and Glymour [**9**] have been able to translate the report into modern terms with certainty.

All three data sets were analyzed and reduced to mean deflection and probable error of the mean. One of the samples was abandoned because it differed considerably from the other two and had a high probable error. The other two were reported separately rather than combined.

It seems clear that neither the authors of the report, nor their audience had a clear view of how data should be combined. Now it would be second nature to combine the samples by weighting each mean by the reciprocal of its variance and then divide by the sum of the weights. This is the linear combination of the constituent means with the lowest variance. This variance is the reciprocal of the sum of the weights.

[1] This was published when scientists waited to publish, until they had something worth publishing.

The following table gives the means and standard deviations of the means for all the samples and the mean and standard deviation of the properly combined samples.

	μ	σ
Sorbal 13-inch astrographic	0.86″	0.48″
Sorbal 4-inch	1.98″	0.178″
Princpe astrographic	1.61″	0.444″
Combined	1.81556″	0.156222″

TABLE 19.1. Deflection Measurements

It has been suggested that Eddington ignored the deviant data from the larger Sorbal telescope because it was least supportive of General Relativity, the theory he wanted to be confirmed. However, Chamont Wang has pointed out that when all three samples are properly combined, the result is nearer the General Relativity prediction than when the deviant sample is discarded [**52**].

Our task is not argue the history, but to answer this question: If the samples were properly combined and the three hypotheses listed in the report were the only choices, what kind of probability statement could have been made about the truth of one of the hypotheses based one the evidence presented.

Figure 19.1 shows the situation. The choice is between three normal distributions, each centered on one of the theoretical values and each with the standard deviation equal to the standard deviation of the combined sample. This is because the error is all instrumental measurement error, the same no matter which hypothesis is true.[2]

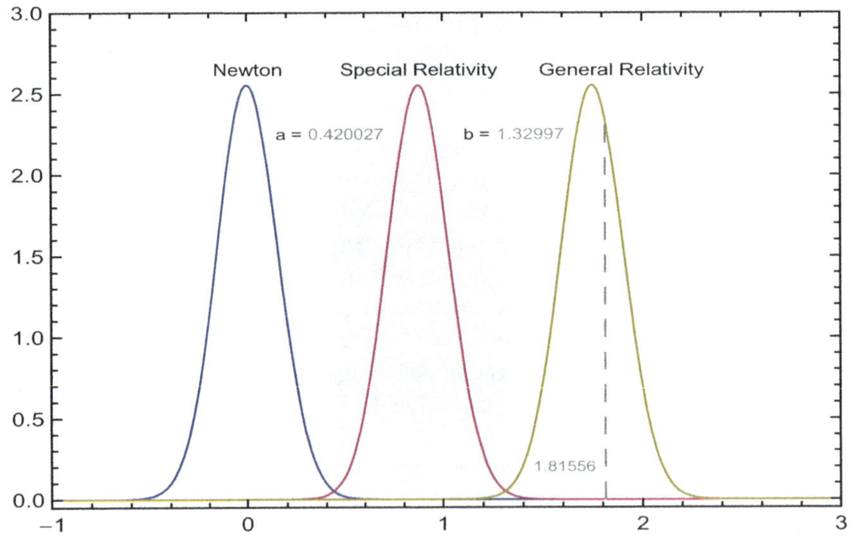

FIGURE 19.1. Eddington's Choice

[2]This fortuitous occurance is almost always true in the hard sciences and almost never true in the soft.

Given this situation, having learned the lessons implicit in the *Further Tales of Sindbad*, how can we devise a rule that has the highest probability of correctly choosing the true hypothesis from among these three. The answer is to chose two values of the mean deflection, $a < b$ such that the area under the curve between $-\infty$ and a in the "Newton" normal distribuion is equal to the area between a and b in the "Special Relativity" normal distribution, and is also equal to the area between b and ∞, in the "General Relativity" normal distribution. The required values are a deflection of $0.420025''$ and $1.322997''$.[3] Table 19.2 summarizes the results of a rule.

	Deflection		
	$-\infty \leftrightarrow 0.420025''$	$0.420025'' \leftrightarrow 1.32997''$	$1.32997'' \leftrightarrow \infty$
Newton: $\mu = 0''$	.996413	.003587	$8.44874 \cdot 10^{-18}$
Special Relativity: $\mu = .875''$	.0017935	.996413	.0017935
General Relativity: $\mu = 1.75''$	$8.44874 \cdot 10^{-18}$	.003587	.996413
	say: "Newton"	say: "Special Relativity"	say: "General Relativity"
	The Decision		

TABLE 19.2. Eddington's Choice

No matter which of the hypotheses is true, there is a .996413 probability that by using the optimum rule, one will correctly identify the true theory—odds of 277.8 to 1, exceeding the 160 to 1 standard extablished by the 5σ rule. If any other decision rule were followed, by increasing the probability of one hypothesis being chosen, one would decrease the probability of another being chosen. Thus, one would have a lower probability of being correct across all three hypotheses.

[3]Looking at Figure 19.1, one might think that the optimum decision points, a and b are midway between the adjacent hypothesis means. This is not true. There is a need to compensate for the fact that the middle distribution loses two tails to the adjacent hypotheses, while the outside two, lose only one.

CHAPTER 20

Stock Market Returns

Since the publication of *Common Stocks as Long-Term Investments* [**45**] helped fuel the great bull market of the 1920s, it has increasingly become a matter of faith that buy and hold in the stock market produces a high rate of return. In the sixties, evidence began to accumulate that, for the most part, stocks followed a random walk with an upward drift [**6**]. Later, with various efficient capital market theories, it has become a matter of *academic* faith, that the way to invest in stocks is to assess one's risk and then simply to buy the proportion of common stocks that matches one's risk preference. After that one just holds on for the long haul. Implicit in all this is, that stocks *have* a positive return. The number usually quoted is nine percent per year, following the University of Chicago study [**14**].

We will examine this question using one hundred ninety-six years of data from 1789 through 1985.[1] We frame the question slightly differently than most. The question will focus on the extent that the total return from stocks excedes inflation. To simplfy notation and formula, we will henceforth assume that the nominal annual return of stocks with reinvested dividends, r_{stocks}, and annual inflation rate, $r_{\text{inflation}}$ have been transformed to a continuously compounded real rate of return, $r_{\text{real}} = log_e(1 + r_{\text{stocks}}) - log_e(1 + r_{\text{inflation}})$.

A graph of this transformed data expressed as a percent is shown in Figure 20.1. The graph shows the year by year, continuously compounded real total return of stocks from 1795 to 1985, along with a thick shaded line equal to the mean return and two thinner shaded lines representing the mean, plus and minus, one standard deviation.

[1]The data is from Ibbotson and Brinson [**23**, chapters 5 and 11]

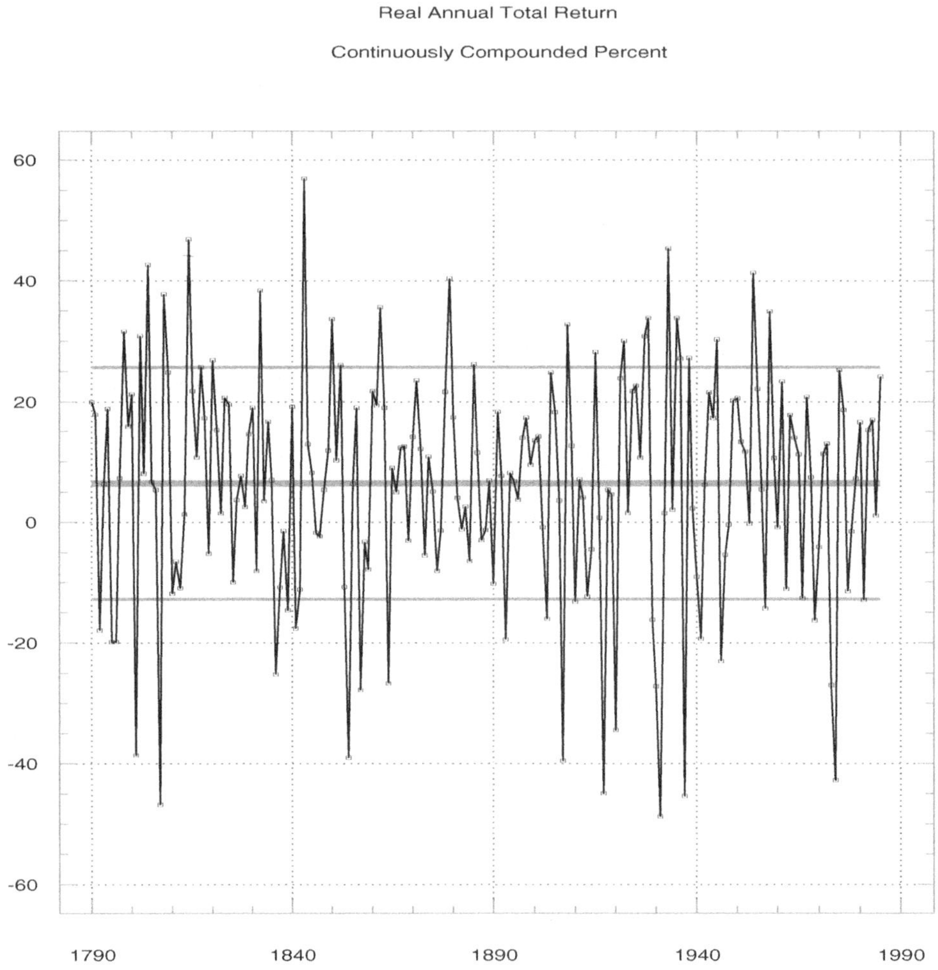

FIGURE 20.1. Real Annual Total Return

Specifically, we will compare two hypotheses. The first is that real yearly total return for the United States stocks follow a serially independent normally distributed random walk with drift equal to the mean of the 196 year data sample and standard deviation equal to the *sample* standard deviation. The second is that they follow a normal distribution with mean zero or less, but with standard deviation unspecified.

Symbolically, the first is that the return is a serially independent normal variate with:

$$\mu = m = m_{1790\text{–}1985} = 0.064101$$
$$\sigma = s = s_{1790\text{–}1985} = 0.192205$$

In accordance with Postulate III, we let a zero mean represent the set of all normal distributions with $\mu \le 0$. Then the second hypothesis is that the rate of return is a normal variate with:

$$\mu = 0$$
$$0 < \sigma < \infty$$

If the first hypothesis holds, one would expect that the standard deviation of the sample mean will shrink in proportion to the reciprocal of the square root of the sample size:

$$s_{m_{1790\text{–}1985}} = \frac{s_{1790\text{–}1985}}{\sqrt{1985-1789}} = \frac{s_{1790\text{–}1985}}{\sqrt{196}}$$
$$= \frac{.192205}{14} = .013729$$

Thus the *first* hypothesis reduces to:

(1μ) $$\mu = 0.064101$$

(1σ) $$\sigma_\mu = s_m = s_{m_{1790\text{–}1985}} = .013729$$

What can we say about the *family of null hypotheses*? Not anymore than we did before!

$$\mu = 0$$
$$0 < \sigma_\mu < \infty$$

Following the principles laid down in Chapter 6 and expanded in Chapter 12, if we are given that $\mu = 0$ and a *sample* has mean, m, and the standard deviation of the sample mean is s_m, then we choose the standard deviation most likely to have produced the sample as a representative of the family,

$$\sigma_\mu = \sqrt{m^2 + s_m^2}$$

This is equivalent to the root square mean square error of the sample measured from zero. *This is the value we can use to devise the rule.*

The *second* hypothesis is:

(2μ) $$\mu = 0$$

(2σ) $$\sigma_\mu = \sqrt{.064010^2 + .013729^2} = .065555$$

With the historic average, (1μ) and (1σ), and the null hypothesis, (2μ) and (2σ), we can formulate a rule such that we can say, "If we follow this rule in deciding whether the target hypothesis is true or null hypothesis is true, we will have a probability $p = .85204$ of being correct. This is true whether the target hypothesis is true or the null hypotheses is true. The value p is correct, no matter which one is true!

Figure 20.2 shows the optimum decision rule.

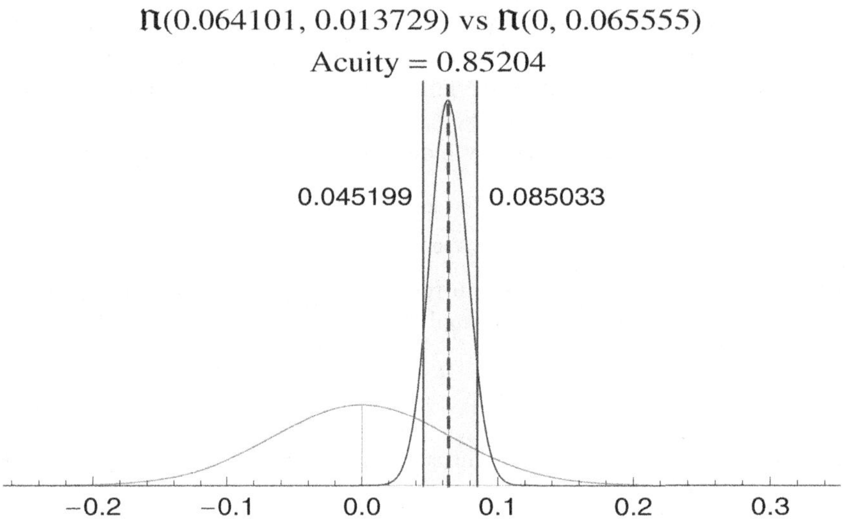

FIGURE 20.2. Historical Performance vs. Zero Return

The tall and narrow normal distribution is the probability distribution of the sample mean for a sample of size 196, if the mean and standard deviation are equal to those that occurred in the years, 1790–1985. The wider normal distribution is probability distribution of the sample mean for a sample of size 196, if the mean is zero and the standard deviation is best aligned with years, 1790–1985 *per* Postulate III.

Table 20.1 shows the mechanics of the decision rule.

	Real Stock Return Sample distribution vs. $\mu = 0$	
	0.045199 $<$ m $>$ **0.085033**	m $<$ **0.045199** or m $>$ **0.085033**
$\mathfrak{h}_1$: Real Return Like 1790–1985	.85204	.14796
$\mathfrak{h}_2$: Zero Real Return	.14796	.85204
	say: "$\mathfrak{h}_1$ is true"	say: "$\mathfrak{h}_2$ is true"
	The Decision: Acuity = .85204	

TABLE 20.1. Real Stock Return

What kind of number is produced by the procedure everyone learns in Statistics 101? The situation is a natural for Student's t-test. The formula is:

$$t_{n-1} = \frac{m}{\frac{s}{\sqrt{n}}}$$

where m is the sample mean, s is the sample standard deviation, and n is the sample size. This statistic is said to follow Student's t-distribution with $n-1$ degrees of freedom. In our case, the formula works itself out like this.

$$\begin{aligned} t_{195} &= \frac{.064101}{\frac{.192205}{\sqrt{196}}} \\ &= 4.66905 \end{aligned}$$

Before computers, one would then have gone to a table to the t-distribution, looked in the row for 195 degrees of freedom, and proclaimed the result "not significant" or "significant at the ... level!" The result presents a slight problem because it is off the table. Mine stops at 3.373, equivalent to being significant at the .0005 level with 120 degrees of freedom. In fact, our result could be considered significant at the 0.00000281 level.

In the past, computational necessity and professional custom joined forces to limit the cutoff points to less extreme levels. The usual choices are significance at the 5%, 1%, and 0.1% levels. To quote a standard textbook:

> Generally, values of α greater than .05 are not considered to be *statistically significant*. A significance level of 5% ($P = 0.05$) corresponds to one type I error in 20 trials, a level of 1% ($P = 0.01$) of one error in 100 trials. Significance of levels of less than 1% ($P = 0.01$) are nearly always adjudged significant; those between 5% and 1% may be considered significant at the discretion of the investigator. [**46**][p. 164]

The authors might have added that $P = 0.001$ is often considered "highly significant." Later they continue explaining the social conventions of their field:

> A convention often encountered is the use of asterisks after the computed value of the significance test, $t_s = 2.86^{**}$. The symbols generally represent the following probability ranges:
>
> $$* = 0.05 \geq P > 0.01 \quad ** = 0.01 \geq P > 0.001 \quad *** = P \leq 0.001$$
>
> [**46**][p. 172]

It is generally regarded as a lapse of good taste to claim significance at a level higher than 0.001—three asterisks will do. Extremely high levels of certainty on this question are regarded as implausible. Paying too close attention to the percentile actually achieved is considered skating on thin ice. If one does not pay close attention, one might mistake the significance level for the probability that the tested hypothesis is true.[2]

Let us imagine a scientist who does decide to *use* 0.00000281 as the level for Type I error.[3] In such a case he would be able state that the odds against such error are greater than 355,751 to 1. Would he not be surprised to find that this result is mathematically equivalent to odds of 5.76 to 1 that his hypothesis is *true*, rather than one of the null hypotheses?

One strategy for increasing the discriminating power of a test is to increase the sample size. Here too there are surprises. If he desires an acuity of the same magnitude as acceptable significance levels there is a fearsome price to pay. In order to achieve acuities of 95%, 99%, and 99.9%, the required sample sizes are 3,232 years, 139,706 years, and 22,799,023 years. Compare these to the 196 years in the current sample.

This result occurs because as data is added, two things occur. The first is that the standard deviation of the sample mean shrinks in proportion to the square root of the number of data points. This is no surprise, but the second thing that happens is! As more evidence is collected, the member of the set of null hypotheses that best fits the expanded data shifts as well. This shift is almost able to keep up with the effect that more data has on the likelihood of the sample, so that there is little change in the probability of the target hypothesis being true, given the increased evidence. The details were spelled out in Chapter 12.

[2]It is commonplace to hear, "There is only a 1% probability that these results were caused by chance alone." To me this seems even closer to relabeling significance level as probability of truth of the hypothesis proposed.

[3]Type I error is rejecting the "null" hypothesis when it is true. Type II error is accepting the "null" hypothesis when it is false.

CHAPTER 21

The Bell Curve

In *The Bell Curve* [**20**], Herrnstein and Murray put forth the argument that heredity largely determines intelligence, as measured by standardized tests, and measured intelligence largely determines success in life. It further argued that forces are at work that will cause a bifurcation between an intelligent successful elite and a mass of hereditarily challenged unsuccessful dullards.

If the authors had not included policy prescriptions and pointed out that race is genetic, the book would have produced little stir. As it turned out, they produced more rancor than can easily be imagined for a book assumed to be a dispassionate sociological treatise. Opinions differed widely as to whether the authors were Hitlerian or merely conservative.

Here will examine the part of their argument that they, most likely, would consider beyond question and, indeed, was mostly unquestioned in the subsequent controversy. That part is the second link in their argument that intelligence measured by standardized tests is a an important predictor of success in life.

Perhaps because they expected their work to be controversial, the authors included a large amount of raw regression results to buttress their opinions. I want to focus on two, one because of all the equations listed as basic analyses, it makes the strongest case; the other because it makes the weakest case—almost no case at all.

The first equation [**20**, p. 597] calculates the probability of permanently dropping out of high school as a function of intelligence, socioeconomic status, and age. The authors use the standard chi-square significance test, a measure of how well the data fit the model.

In this case, the model chi-square is 787.7956 with three degrees of freedom, $\chi^2_{[3]} = 787.7956$. The authors, if I read them right seem to indicate that this result is significant at the level of .0000005. The authors are entirely too modest! The actual number is 1.92×10^{-170}, or 192 preceded by a decimal point and 169 zeros. Expressed as odds, the result is 5.21×10^{169} to one.

The ratio between size a quark and the size of the universe is about 10^{42} [**36**]. If our universe were a quark in a larger universe and that universe were a quark in a third universe and that universe were a quark in a fourth universe and there were *something* fifty times bigger than that fourth universe, then the size of that *something* would be 5×10^{169} bigger than our quark.

If one could properly interpret this result as the odds that their model is true rather than that there is simply *average* probability of dropping out of high school, independent of measured intelligence, socioeconomic status, or age; the weight of this evidence would be astronomical, one might even say that it was astronomical to the fourth power!

Unfortunately, the calculation is the answer to another question. It is the answer to, "What is the probability of drawing a sample such that the values of measured intelligence, socioeconomic status, and age take these values and in combination with the equation presented one can forcast 'Permanently Dropping Out of High School' with this accuracy if, *in fact* there is zero correlation between the independent variables and the dependent variable?"

If one wishes to choose correctly between the model presented and the null hypothesis that there is just an *average probability* of dropping out of high school, the answers are literally light-years apart.

As explicated in Chapter 17, for large n, the χ^2 distribution converges to the normal distribution. Partitioning the sum of squared error between that explained by the mean and the sum of the squared error explained by the regression equation, the decision reduces to the choice between two normal distributions, one with mean $= n - 1$ and standard deviation $\sqrt{2(n-1)}$, the other with mean $= (n-1)(1-\overline{R}^2)$ and standard deviation $= \sqrt{2(n-k-1)}$ where k is the number of explanatory variables.

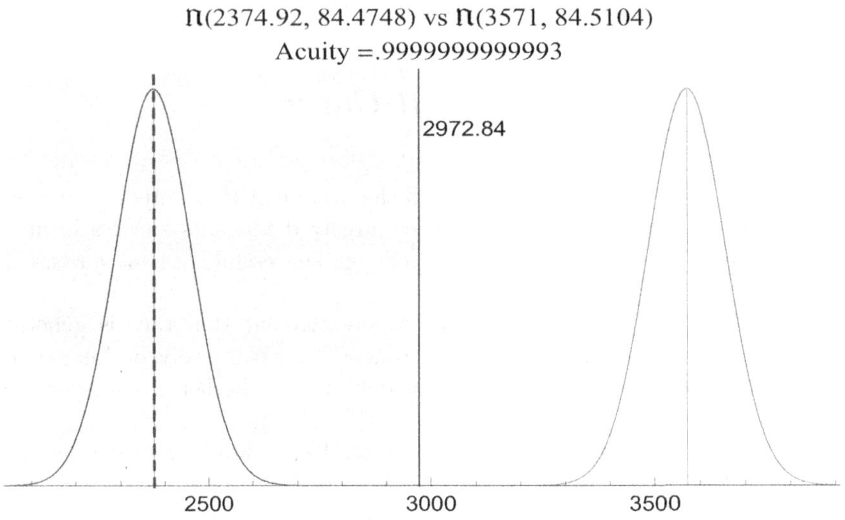

FIGURE 21.1. Permanently Dropped Out of High School

	Permanently Dropped Out of High School Intelligence, Socioeconomic Status, and Age	
	$m < 2972.84$	$m > 2972.84$
$\mathfrak{h}_1$: Tendency to have permanently dropped out of high school is determined by intelligence, socioeconomic status, and age	.9999999999993	.0000000000007
$\mathfrak{h}_2$: Tendency to have permanently dropped out of high school is is just a normally distributed random variable	.0000000000007	.9999999999993
	say: "$\mathfrak{h}_1$ is true"	**say: "$\mathfrak{h}_2$ is true"**
	The Decision: Acuity = .9999999999993	

TABLE 21.1. Permanently Dropped Out of High School

The same sample evidence allows the experimenter to make two statements:

(1) The odds that the sample of independent and dependent variables was drawn from a world in which "Permanently Dropping Out of High School" *is not explained by measured intelligence, socioeconomic status, and age*—only by an average tendency $= 5.211 \times 10^{169}$ to 1.

(2) If the sample was drawn from a world in which "Permanently Dropping Out of High School" is explained only by an average tendency *or* a world where the regression model is *true*, *the odds that the regression is true* = 1.3674×10^{12} to 1, approximately the ratio of the size of a quark to the diameter of a red blood cell.

This is a very strong result. It far, far exceeds the threshold of 160 to 1 that the 5σ rule has tacitly established for the hard sciences.

On reflection, one should not be too surprised at the result. Intelligence, the main driver of the model, is the basic currency of schooling. One would not be surprised if someone with an aptitude for chess, who won a lot and enjoyed it, continued to play it. This is close as one could get to a tautology in the social sciences. Still, putting the actual number down for all to see claims an embarrassingly high level of certainty.

Herrnstein and Murray extend the argument that intelligence is the driver of social ills beyond the point that, looked at in the right way, the evidence presented supports it. The weakest equation is the one presented on page 601. It purports to explain "Ever Married before Age of 30" with the same variables, measured intelligence, socioeconomic status, and age. All seems to be well. The equation is significant at the $p = .005$ level. The when the significance level is translated into odds, the result is 201.7 to 1.

In fact, if ones asks the standard question, "If the model is true or 'Ever Married before Age of 30' is just fluctuating around a mean is true, given the evidence, using the best decision rule available, what is the probability of deciding correctly in favor of the model?" the answer is pathetically weak as shown in Figure 21.2 and Table 21.2. The probability of correctly coming down on the side of the authors' equation is only 0.53289 with odds of 1.1408 to 1. An hypothesis with basically no evidential support is seen by significance testing as a well substantiated result.

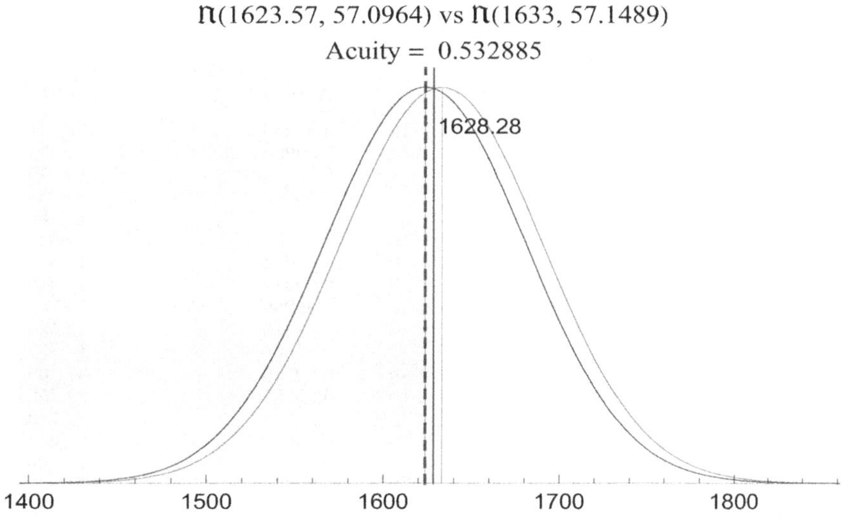

FIGURE 21.2. Ever Married Before Age of 30

	Ever Married Before Age of 30 Intelligence, Socioeconomic Status, and Age	
	$m < 1628.28$	$m > 1628.28$
$\mathfrak{h}_1$: Tendency to have ever married before age of 30 is determined by intelligence, socioeconomic status, and age	.532885	.467115
$\mathfrak{h}_2$: Tendency to have ever married before age of 30 is just a normally distributed random variable	.467115	.532885
	say: "$\mathfrak{h}_1$ is true"	**say: "$\mathfrak{h}_2$ is true"**
	The Decision: Acuity = .532885	

TABLE 21.2. Ever Married Before Age of 30

Perhaps because of the low explanatory power the model, perhaps because intelligence contributes almost no explanatory power, there is no mention of this result in the body of the book. It is completely relegated to the appendix.

Herrnstein and Murray present eighteen main equations, summarized in Table 21.3. Of these sixteen are significant at the level of $p = .001$, earning them a ***. The other two are significant at the level of $p = .01$, earning them **.

When one asks the correct question, the results are much different. Assume the probability of deciding truth incorrectly was awarded stars in the standard way: $p \leq .05$ gets *, $p \leq .01$ gets **, and $p \leq .001$ gets ***. Using Acuity in the same way, twelve of the equations would get no stars, two would get *, two would get ** and two would get ***. Only three meet the criterion that would be expected in the hard sciences, odds of 160 to 1. Nine of the 18 fail the Russian roulette threshold of 5 to 1 odds of success.

Bell Curve Equations							
		Significance			Optimum Decision		
No.	Page	Level		Odds	Acuity		Odds
1	595	3.4627×10^{-39}	***	2.8879×10^{38}	0.98280	*	57.131
2	597	1.9189×10^{-170}	***	5.2113×10^{169}	0.9999999999993	***	1.3674×10^{12}
3	599	0.00028866	***	3463.2	0.58703		1.4215
4	600	0.000048533	***	20603.	0.65079		1.8636
5	601	0.0049335	**	201.70	0.53289		1.1408
6	602	7.1403×10^{-9}	***	1.4005×10^{8}	0.62303		1.6528
7	604	3.4946×10^{-17}	***	2.8615×10^{16}	0.83004		4.8837
8	607	1.9537×10^{-41}	***	5.1186×10^{40}	0.99919	***	1230.5
9	608	7.9484×10^{-18}	***	1.2581×10^{17}	0.99280	**	137.85
10	609	1.7559×10^{-36}	***	5.6952×10^{35}	0.82310		4.6530
11	609	0.0044171	**	225.39	0.61356		1.5877
12	611	2.1436×10^{-34}	***	4.6650×10^{33}	0.99736	**	377.09
13	612	5.5542×10^{-35}	***	1.8004×10^{34}	0.95789	*	22.750
14	615	6.3429×10^{-14}	***	1.5766×10^{13}	0.70338		2.3713
15	617	6.2241×10^{-9}	***	1.6067×10^{8}	0.83628		5.1078
16	620	0.00016537	***	6046.1	0.58608		1.4160
17	621	4.1684×10^{-10}	***	2.3990×10^{9}	0.92972		13.229
18	622	8.4493×10^{-70}	***	1.1835×10^{69}	0.93064		13.418

TABLE 21.3. Summary of Bell Curve Results

Table 21.4 shows the six that have some support:

Bell Curve Equations			
No.	**Page**		**Hypothesis**
1	595	*	Under the Official Poverty Line in 1989
2	597	***	Permanently Dropped Out of High School
8	607	***	On Welfare by the First Calendar Year After the Birth of the Child
9	608	**	On Welfare for at Least Five Years vs. Women with No Welfare Experience
12	611	**	The Child's Mother Was Under the Poverty Line Though-out the Child's First 3 Years of Life
13	612	*	The Child's HOME Index Score Was in the Bottom Decile

TABLE 21.4. Bell Curve Results with Some Evidence of Being True

The grand scheme of explaining everything on the basis of intelligence is in shambles.[1]

The entire controversy falls back to the question of whether intelligence measured by standardized tests is largely determined by heredity and as an ancillary question, whether race is an important determinant—a question we do not address.

[1]On page 594 the authors make this statement: "This illustrates why we therefore consider regression coefficients themselves (and their associated key values) to suit our analytic purposes better than R^2 and that is why those are the ones we relied on in the text." This is a conceptual blunder. The authors present smooth curves backed out of logit space in the illustrations in the body of the book. This procedure gives no weight to the explanatory power of the total model. Before one draws conclusions from a model, one must first consider the question of whether the model is true. Only then can one draw conclusions from parts of the model.

CHAPTER 22

Laffer & Ranson: A Formal Model of the Economy

In 1971, Arthur B. Laffer and R. David Ranson published *A Formal Model of the Economy* in the The Journal of Business. [**29**] It was controversial for two reasons. The first was that rather than using seasonally adjusted data, they used raw data and did the seasonal adjustment in the context of estimating the model. The second was that their model showed a number of things that were contrary to the received wisdom of the economics profession. For instance, there was no multiplier effect for government spending. Government spending increased GNP in the quarter that ended occurred but was erased by lowered GNP in the succeeding three quarters. There was no lagged effect of monetary policy. An increase in money supply, M1, by 1% permanently increased GNP by 1%, but had no effect afterwards.

One is not so much concerned about the economic theory implications of the model, but rather, whether there was a successfully estimated economic model at all. Because the equations in the Laffer/Ranson model[1] are changes in trending variables or non-trending variables it is possible to use $\overline{R}^2$ to test the results against "just fluctuating around a mean."

There are seemingly successful explanations of change in nominal GNP, real GNP, and the unemployment rate. The explanations of changes in the GNP deflator proved to be below the standard that is acceptable.

22.1. Nominal GNP

The first equation, that for nominal GNP, is quite successful against *this* straw man. As shown in Figure 22.1 there is a .999434 probability of correctly choosing the model over the idea that the rate of change in GNP is just fluctuating around a mean. The odds of choosing correctly are 1765.78 to one. If one incorrectly calculated the odds using one minus the significance level the result would be 1.097310×10^{50} to one. This shows why practitioners in the soft sciences regularly overvalue their results.

None the less, the Laffer model is decisive against "just fluctuates around a mean."

[1]Henceforth, I will refer to the Laffer model because of his distinguished career in napkin decoration for the distinguished thinkers, Donald Rumsfeld and Dick Cheney.
http://web.archive.org/web/20110503200219/http://www.polyconomics.com/gallery/Napkin003.jpg

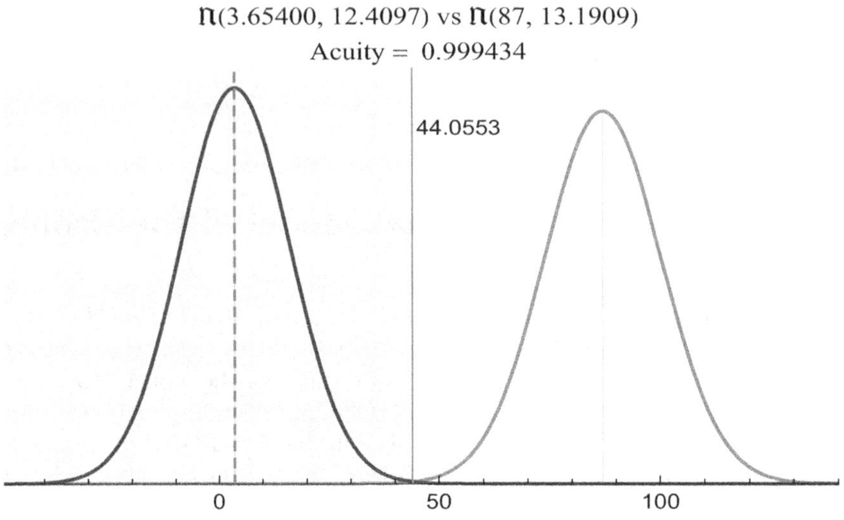

FIGURE 22.1. Nominal GNP: 1948:I–1969:IV using M1, $\overline{R}^2 = .958$

	Change in Nominal GNP Seasonal Factors, Changes in M1, Strike Activity, Stock Prices, and Current and Lagged Changes in Government Spending	
	$m < 44.0553$	$m > 44.0553$
$\mathfrak{h}_1$: The Laffer Model is correct.	.999434	.000566
$\mathfrak{h}_2$: Changes in nominal GNP are just a random normal variate.	.000566	.999434
	say: "$\mathfrak{h}_1$ is true"	**say: "$\mathfrak{h}_2$ is true"**
	The Decision: Acuity = .999434	

TABLE 22.1. Change in Nominal GNP

One might think that, that's the end of the story. However it's notable that a great deal of the explained variance in the Laffer model is explained by dummy variables for seasonality. When one uses *just* seasonality dummy variables, almost as much of the variation is explained. The seasonality model is a more apt "straw man" than "just fluctuating around a mean." Since Laffer has chosen to incorporate seasonal variables into his overall model when others use seasonally adjusted variables, it seems rather unfair to claim credit for seasonality when others are evaluating their models as an explanation above and beyond seasonality. The seemingly positive result is muted by the fact that most of the variation is captured by seasonality alone. Just regressing $\Delta \text{Ln(GNP)}$ on seasonal dummy variables yields an $\overline{R}^2$ of 0.898 and an acuity of .998586.

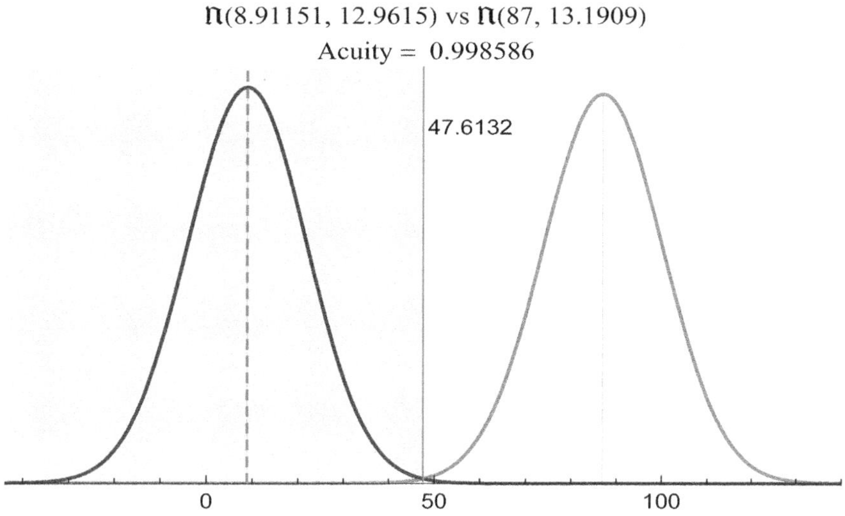

FIGURE 22.2. Nominal GNP: 1948:I–1969:IV Seasonal, $\overline{R}^2 = .898$

	Change in Nominal GNP Seasonal Dummy Variables	
	$m < \mathbf{47.6132}$	$m > \mathbf{47.6132}$
$\mathfrak{h}_1$: Seasonal adjustment is correct.	.998586	.001414
$\mathfrak{h}_2$: Changes in nominal GNP are just a random normal variate.	.001414	.998586
	say: "$\mathfrak{h}_1$ is true"	**say: "$\mathfrak{h}_2$ is true"**
	The Decision: Acuity = .998586	

TABLE 22.2. Change in Nominal GNP

Since the distributions representing the Laffer and Seasonal models are located in relationship to the location of the distribution for "just fluctuating around a mean" for each of the regressands, the relative positions of the two models' distributions can be determined.

When the the full model is pitted against "Only Seasonality," the full model is favored with an acuitly of .58 or odds of 1.4 to 1 in its favor—not a rousing recommendation.[2]

22.2. Inflation

There are two equations for inflation. The first leaves out the difficult period from 1948 to 1952. However, neither comes up to an adequate standard. In Figure 22.3 and Table 22.3 the acuity is just .909693. These translate into the odds

[2]The numbers are approximate and there are no graphs because the normal approximations to the χ^2 distributions so close to zero and close to each other.

of choosing correctly of 10.0733 to 1. If significance levels were used to calculate odds the result would be 1.348421×10^8 to 1.

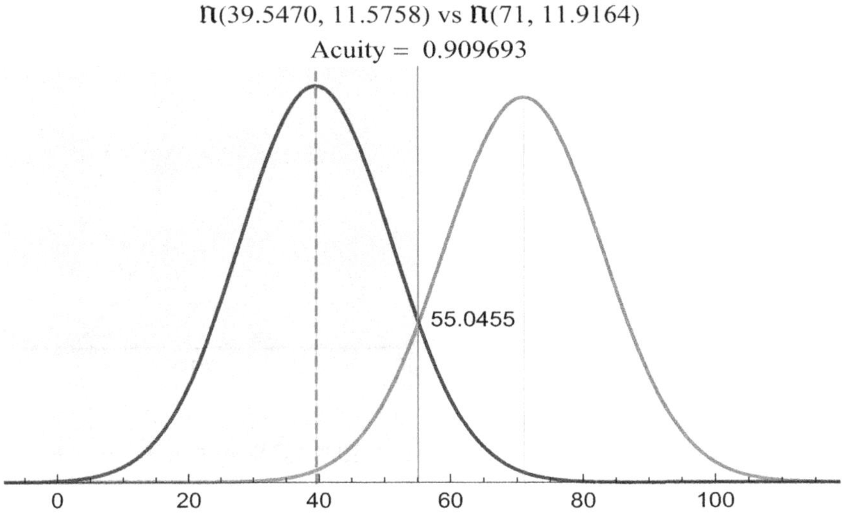

FIGURE 22.3. Inflation: 1952:I–1969:IV, $\overline{R}^2 = .443$

	Change in GNP Deflator Changes in M1, Lagged Changes in GNP Deflator and the t-Bill Rate	
	$m < 55.0455$	$m > 55.0455$
$\mathfrak{h}_1$: The Laffer Model is correct.	.9099693	.090307
$\mathfrak{h}_2$: Changes in GNP Deflator are just a random normal variate.	.090307	.9099693
	say: "$\mathfrak{h}_1$ is true"	**say: "$\mathfrak{h}_2$ is true"**
	The Decision: Acuity = .9099693	

TABLE 22.3. Change in GNP Deflator

When the war influenced years are included, in Figure 22.4 and Table 22.4, the acuity is just .883192 and odd of 7.56105 to 1. If significance levels were used to calculate odds the result would be 820,994.3 to 1.

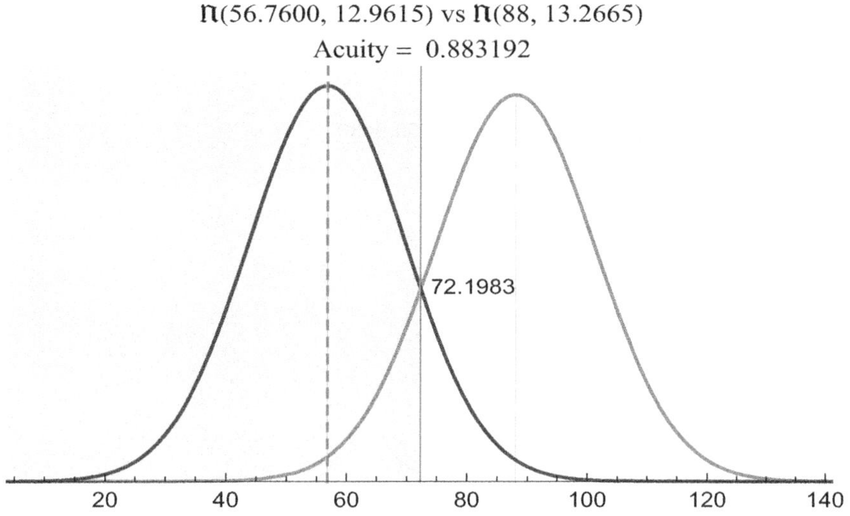

FIGURE 22.4. Inflation: 1948:I–1969:IV, $\overline{R}^2 = .355$

	Change in GNP Deflator, 1948–1951 Included Changes in M1, Lagged Changes in GNP Deflator and the t-Bill Rate	
	$\boldsymbol{m < 72.1983}$	$\boldsymbol{m > 72.1983}$
$\mathfrak{h}_1$: The Model is correct.	.883192	.116808
$\mathfrak{h}_2$: Changes in GNP Deflator are just a random normal variate.	.116808	.883192
	say: "$\mathfrak{h}_1$ is true"	**say: "$\mathfrak{h}_2$ is true"**
	The Decision: Acuity = .883192	

TABLE 22.4. Change in GNP Deflator, 1948–1951 Included

During the non-war years, there was little inflation to be explained and when there was inflation, it was not adequately explained.

22.3. Unemployment

Figure 22.5 and Table 22.5 show the decision rule for their equation predicting the unemployment rate. The acuity is .996267, which translates into the odds of being correct of 266.865 to one, contrasted to 4.773956×10^{25} to one, if derived from the significance level.

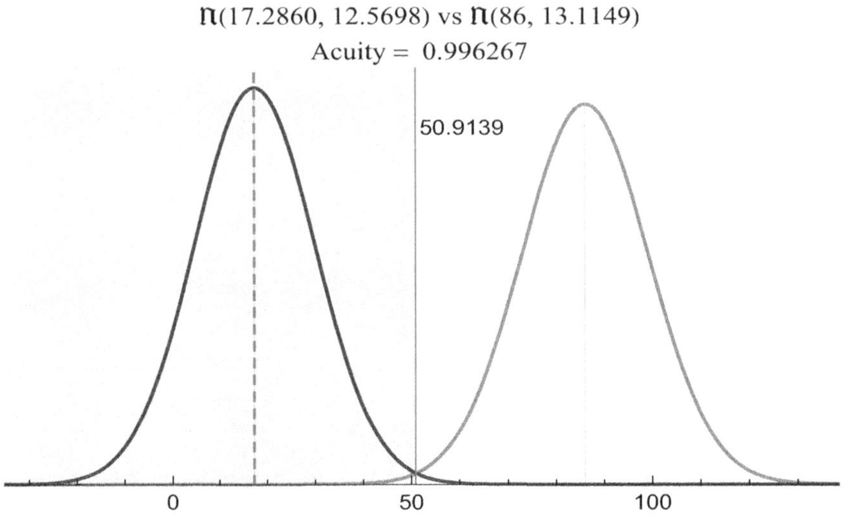

FIGURE 22.5. Unemployment: 1948:II–1969:IV, $\overline{R}^2 = .799$

	Change in the Unemployment Rate Seasonal Factors and Current and Lagged Changes in Real GNP	
	$m < \mathbf{50.91397}$	$m > \mathbf{50.91397}$
$\mathfrak{h}_1$: The Laffer Model is correct.	.996267	.003733
$\mathfrak{h}_2$: Changes in the Unemployment Rate are just a random normal variate.	.003733	.996267
	say: **"$\mathfrak{h}_1$ is true"**	**say:** **"$\mathfrak{h}_2$ is true"**
	The Decision: Acuity = .996267	

TABLE 22.5. Change in the Unemployment Rate

All would seem well, but this equation does not stand up so well when paired with an equation that just uses dummy seasonal variables. Figure 22.6 shows the results of "just seasonal variables" versus "just fluctuating around a mean."

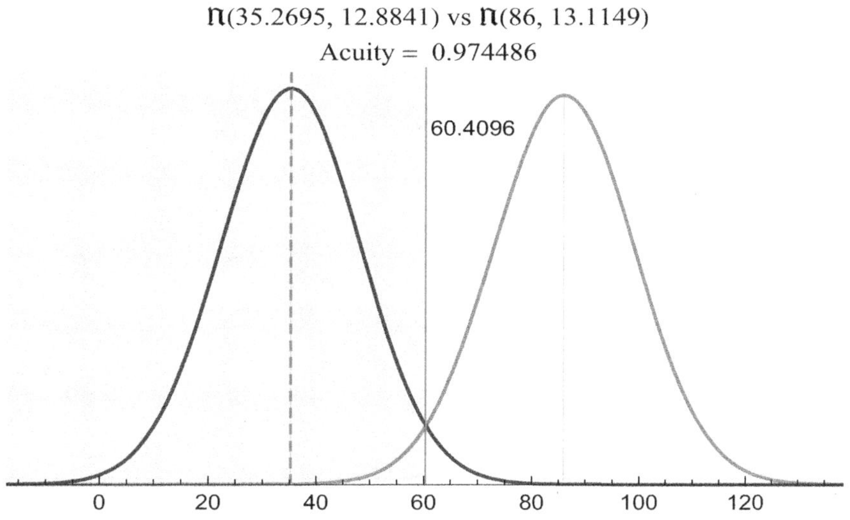

FIGURE 22.6. Unemployment: 1948:II–1969:IV, Seasonal $\overline{R}^2 = .590$

	Change in the Unemployment Rate Only Seasonal Factors	
	$m < 60.4096$	$m > 60.4096$
$\mathfrak{h}_1$: Seasonal Dummy Variables	.974486	.025514
$\mathfrak{h}_2$: Changes in the Unemployment Rate are just a random normal variate.	.025514	.974486
	say: "$\mathfrak{h}_1$ is true"	**say: "$\mathfrak{h}_2$ is true"**
	The Decision: Acuity = .974486	

TABLE 22.6. Change in the Unemployment Rate

When the the full model is pitted against "Only Seasonality," the full model is favored with an acuitly of .56 or odds of 1.3 to 1 in its favor.

22.4. Real GNP

Real GNP, Figure 22.7 and Table 22.7, seems to be a winner. The acuity is, .999481 which translates into odds of 1925.78 to one, contrasted with 3.283663×10^{46} to one, if significance levels are used. However, it too benefits from being put up against too weak straw man—seasonality explains almost all the variation.

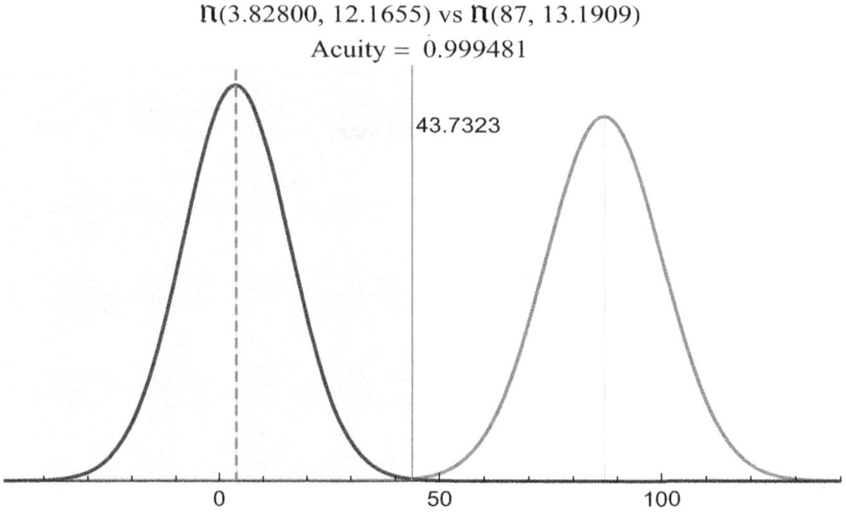

FIGURE 22.7. Real GNP: 1948:I–1969:IV, $\overline{R}^2 = .956$

	Change in Real GNP Seasonal Factors, Current Changes in M1, Strike Activity, Stock Prices, Lagged Changes in GNP Deflator, and Current and Lagged Changes in Government Spending	
	$m < 43.7323$	$m > 43.7323$
$\mathfrak{h}_1$: The Laffer Model is correct.	.999481	.000519
$\mathfrak{h}_2$: Changes in Real GNP are just a random normal variate.	.000519	.999481
	say: "$\mathfrak{h}_1$ is true"	**say: "$\mathfrak{h}_2$ is true"**
	The Decision: Acuity = .999481	

TABLE 22.7. Change in Real GNP

When regressed on seasonal dummy variables alone, Figure 22.8 and Table 22.8, the results are nearly as good. Acuity equals .998753; odds equal 800.92 to 1.

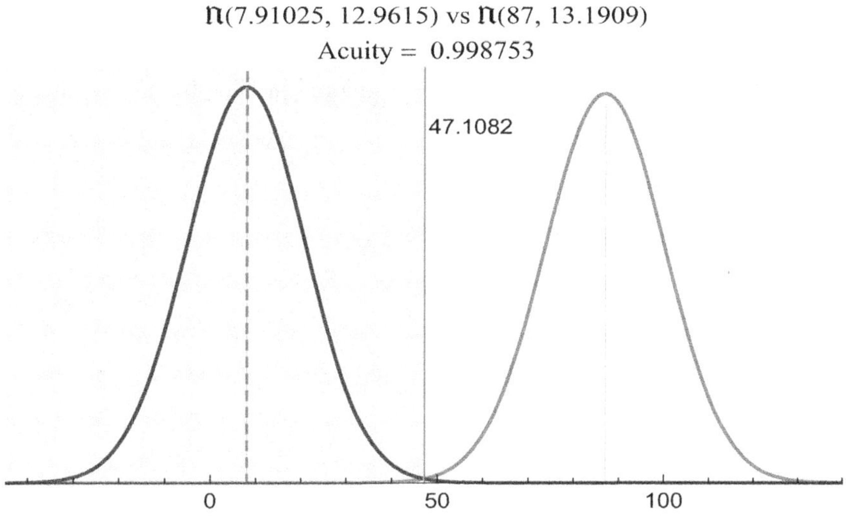

FIGURE 22.8. Real GNP: 1948:I–1969:IV, $\overline{R}^2 = .909$

	Change in Real GNP Seasonal Dummy Variables	
	$m < \mathbf{47.1082}$	$m > \mathbf{47.1082}$
$\mathfrak{h}_1$: Seasonal adjustment is correct.	.998753	.001347
$\mathfrak{h}_2$: Changes in real GNP are just a random normal variate.	.001347	.998753
	say: **"$\mathfrak{h}_1$ is true"**	**say:** **"$\mathfrak{h}_2$ is true"**
	The Decision: Acuity = .998753	

TABLE 22.8. Change in Real GNP

When the the full model is pitted against "Only Seasonality," the full model is favored with an acuity of .76 or odds of 3.2 to 1 in its favor.

So they were successful for changes in nominal GNP, real GNP, and the unemployment rate when measured against, "just fluctuating around a mean." In all three cases, regressing on only seasonal dummy variables gave results that were almost as good.

Both models of the inflation rate, failed to meet standards that would be taken seriously against no model at all.

Since the policy variables did not raise the acuity in a meaningful way, it is hard to think that *any* conclusions can be drawn from Laffer's model. If the policy variables had negligible effect, their coefficients are of no interest.

Acuity depends on the choice of hypotheses and the evidence, and nothing else. The situation calls out for replication of the experiment. To do this one would hope the equations constant and apply them to new data. The exact form of the test would depend on the number of data points in the replication and the mean and standard deviation of the new data.

It is possible that the new data would reinforce the equations that did not meet scientific standards on the initial data; it also may be the case, that the good equations fail to explain data outside their initial range. Sadly, this is rarely the case. Details of the replication of regression equation were explored in Chapter 17.

CHAPTER 23

Bem's Precognition Experiment

In 2011, the *Journal of Personality and Social Psychology* published an article by Daryl J. Bem [**3**] giving evidence for a a variety of paranormal phenomena. Aggregating the results of all the significance tests by means of Stouffer's $z = 6.66$, he produced $p = 1.34 \times 10^{-11}$. Paranormal phenomena are, almost by definition, what clashes with common sense. They are phenomena that the great majority of people do not believe to exist.

Two sets of Bayesian analysts rose to the occasion to beat back what most believed to be erroneous results and simultaneously promote their own method of analysis. The first, by Wagenmakers, *et al.*, found that each individual experiment when looked at through a Bayesian lens was not justified [**50**]; the second, engaging in their own brand of meta-analysis, Rouder and Morey found that there was a there, there, but the there wasn't big enough to overcome appropriate skepticism [**42**]. Bem, *et al.* wrote a rebuttal that by modifying the prior distribution causes Bayesian analysis to come down in favor of PSI [**4**].

Of course, the Bayesian procedures can be adjusted until one can "justifiably" believe whatever one wants to believe. One can even believe something without using Bayesian procedures. It is also true that significance testing is of no use in assigning a probability to the choice between competing hypotheses.

I would argue that belief is irrelevant. The task is to make a true statement based on the forms of the competing hypotheses and the evidence; beliefs should not enter into it.

The first two of Bem's experiments offer a chance to compare hypotheses with different means and known standard deviations. We assume on distribution is a normal approximation to a binomial distribution consistent with Bem's data and that the alternative distribution is a binomial distribution with $p = \frac{1}{2}$. To avoid the possibility of a cut at a negative number or a number greater than n in the case of a two-cut rule, the decision rule is a one-cut rule.

Experiment 1: Precognitive Detection of Erotic Stimuli

"As noted above, most of the earlier experiments in precognition explicitly challenged participants to guess which one of several stimuli would be randomly selected after they recorded their guess. In most of these experiments, participants were also given explicit trial-by-trial feedback on their performance. This first experiment adopts that same traditional protocol, using erotic pictures as explicit reinforcement for correct precognitive guesses.

Method

"One hundred Cornell undergraduates, 50 women and 50 men, were recruited for this experiment using the Psychology Department's automated online sign-up system. They either received one point of experimental credit in a psychology course offering that option or were paid \$5 for their participation. Both the recruiting announcement and the introductory explanation given to participants upon entering the laboratory informed them that

> This is an experiment that tests for ESP. It takes about 20 minutes and is run completely by computer. First you will answer a couple of brief questions. Then, on each trial of the experiment, pictures of two curtains will appear on the screen side by side. One of them has a picture behind it; the other has a blank wall behind it. Your task is to click on the curtain that you feel has the picture behind it. The curtain will then open, permitting you to see if you selected the correct curtain. There will be 36 trials in all. Several of the pictures contain explicit erotic images (e.g., couples engaged in nonviolent but

explicit consensual sexual acts). If you object to seeing such images, you should not participate in this experiment.

"The participant then signed a consent form and was seated in front of the computer. After responding to two individual-difference items (discussed below), the participant had a 3-min relaxation period during which the screen displayed a slowly moving Hubble photograph of the starry sky while peaceful new-age music played through stereo speakers. The 36 trials began immediately after the relaxation period.

"**Stimuli.** Most of the pictures used in this experiment were selected from the International Affective Picture System (IAPS; Lang & Greenwald, 1993), a set of 820 digitized photographs that have been rated on 9-point scales for valence and arousal by both male and female raters. This is the same source of pictures used in most presentiment studies. Each session of the experiment included both erotic and nonerotic pictures randomly intermixed, and the main psi hypothesis was that participants would be able to identify the position of the hidden erotic picture significantly more often than chance (50%).

"The hit rate on erotic trials can also be compared with the hit rates on the nonerotic trials to test whether there is something unique about erotic content in addition to its positive valence and high arousal value. For this purpose, 40 of the sessions comprised 12 trials using erotic pictures, 12 trials using negative pictures, and 12 trials using neutral pictures. The sequencing of the pictures and their left/right positions were randomly determined by the programming language's internal random function. The remaining 60 sessions comprised 18 trials using erotic pictures and 18 trials using nonerotic positive pictures with both high and low arousal ratings. These included eight pictures featuring couples in romantic but nonerotic situations (e.g., a romantic kiss, a bride and groom at their wedding). The sequencing of the pictures on these trials was randomly determined by a randomizing algorithm devised by Marsaglia (1997), and their left/right target positions were determined by an Araneus Alea I hardware-based random number generator. (The rationale for using different randomizing procedures is discussed in detail below.) "Although it is always desirable to have as many trials as possible in an experiment, there are practical constraints limiting the number of critical trials that can be included in this and several others experiments reported in this article. In particular, on all the experiments using highly arousing erotic or negative stimuli a relatively large number of nonarousing trials must be included to permit the participant's arousal level to "settle down" between critical trials. This requires including many trials that do not contribute directly to the effect being tested.

"In our first retroactive experiment (Experiments 5, described below), women showed psi effects to highly arousing stimuli but men did not. Because this appeared to have arisen from men's lower arousal to such stimuli, we introduced different erotic and negative pictures for men and women in subsequent studies, including this one, using stronger and more explicit images from Internet sites for the men. We also provided two additional sets of erotic pictures so that men could choose the option of seeing male-male erotic images and women could choose the option of seeing female-female erotic images.

"From the participants' point of view, this procedure appears to test for clairvoyance. That is, they were told that a picture was hidden behind one of the curtains and their challenge was to guess correctly which curtain concealed the picture. In fact, however, neither the picture itself nor its left/right position was determined until after the participant recorded his or her guess, making the procedure a test of detecting a future event, that is, a test of precognition."

Of all the 1560 erotic pictures, the subjects guessed correctly 53.1% of the time. Using a binomial test Bem calculated that $p = .011$. Redoing the exercise as a choice between binomial distributions with $p = .5$ and $p = .531$, the probability of choosing correctly is .88979 as shown in Figure and Table 23.1.

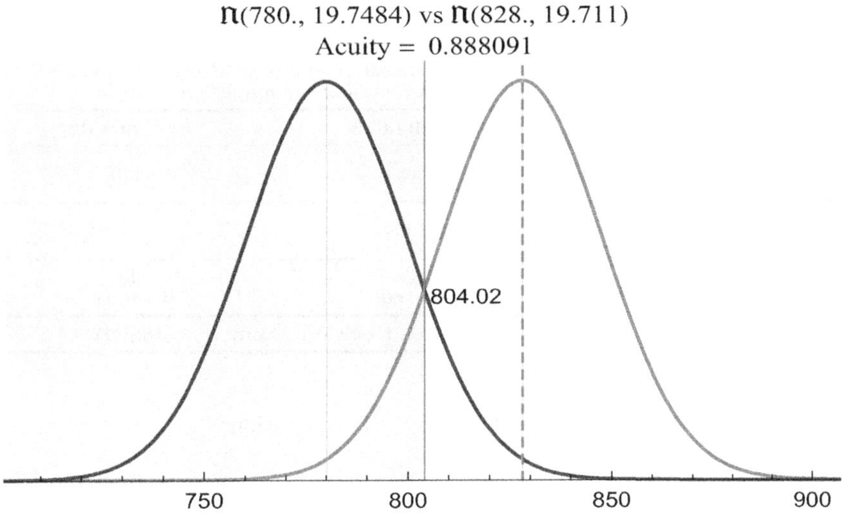

FIGURE 23.1. Erotic Precognition vs. Random

	Erotic Precognition The erotic charge of a picture allows a subject to correctly predict the location of a picture *when the location is chosen after the prediction is made*	
	$m > 804.023$	$m < 804.023$
$\mathfrak{h}_1$: Erotic Precognition is real	.888091	.111909
$\mathfrak{h}_2$: Erotic Precognition is just a statistical artifact	.111909	.888091
	say: "$\mathfrak{h}_1$ is true"	**say: "$\mathfrak{h}_2$ is true"**
	The Decision: Acuity = .888091	

TABLE 23.1. Erotic Precognition

The odds of choosing correctly are 7.93586 to 1, a far cry from the 277 to one odds that we saw in the Eddington expedition results or the 160 to 1 odds that is the tacit standard for the hard sciences or the 90 to 1 odds Bem might have imagined if he misinterpreted the significance level.

Experiment 2: Precognitive Avoidance of Negative Stimuli

Method

"One hundred fifty Cornell undergraduates, 107 women and 43 men, were recruited for this experiment using the Psychology Department's automated online sign-up system. Both the recruiting announcement and the opening instructions given to participants upon entering the laboratory, informed them that

> This is an experiment that tests for ESP (Extrasensory Perception). The experiment is run entirely by computer and takes about 15 minutes. On each trial of the experiment you will be shown a picture and its mirror image side by side and asked to indicate which image you like better. The computer will then flash a masked picture on the screen. The way in which this procedure tests for ESP will be explained to you at the end of the session.

"Note that the participants task in this experiment was simply to express a preference between two closely matched pictures on each trial; unlike in traditional precognition experiments (and Experiment 1 reported above), the participant was not faced with an explicit psi challenge.

"As in Experiment 1, the participant was seated in front of the computer and asked to respond to the two items on the stimulus seeking scale. This was followed by the same 3-min relaxation period. Then, on each of 36 trials, the participant was shown a low-arousal, affectively neutral picture and its mirror image side by side and asked to press one of two keys on the keyboard to indicate which neutral picture he or she liked better. Using the Araneus Alea I hardware-based RNG, the computer then randomly designated one of the two pictures to be the target. Note that the computer did not determine which of the two neutral pictures would be the target until after the participant had registered his or her preference. Whenever the participant had indicated a preference for the target-to-be, the computer flashed a positively valenced picture on the screen subliminally three times. Whenever the participant had indicated a preference for the non-target, the computer subliminally flashed a highly-arousing, negatively valenced picture. Pictures were again selected primarily from the IAPS set.

"The flashed pictures were exposed for 33 ms followed immediately by a masking stimulus for 167 ms. Time between flashes was 500 ms. A Hubble photograph of the starry sky appeared on the screen for 3000 ms before the onset of the next trial. A hit was defined as preferring the target-to-be, the picture that avoided the subliminally exposed negative picture. Because participants chose between two images on each trial, the psi hypothesis was that they would prefer the target to the nontarget on significantly more than 50% of the trials.

"The pairs of neutral pictures were presented in a fixed order for all sessions, and the RNG randomly determined the left/right position of the two images. For the first 100 sessions, the flashed positive and negative pictures were independently selected and sequenced randomly. For the subsequent 50 sessions, the negative pictures were put into a fixed sequence, ranging from those that had been successfully avoided most frequently during the first 100 sessions to those that had been avoided least frequently. If the participant selected the target, the positive picture was flashed subliminally as before, but the unexposed negative picture was retained for the next trial; if the participant selected the nontarget, the negative picture was flashed and the next positive and negative pictures in the queue were used for the next trial. In other words, no picture was exposed more than once, but a successfully avoided negative picture was retained over trials until it was eventually invoked by the participant and exposed subliminally. The working hypothesis behind this variation in the study was that the psi effect might be stronger if the most successfully avoided negative stimuli were used repeatedly until they were eventually invoked."

Across all 5400 trials, 51.7% correctly avoided negative images. Bem calculates $z = 2.44$ and $p = .007$. Conjuring up odds from these numbers, one gets 141.9 to one. Deciding between a binomial distribution with $p = .5$ and $p = .517$ with $n = 5400$, is shown in Figure and Table 23.2.

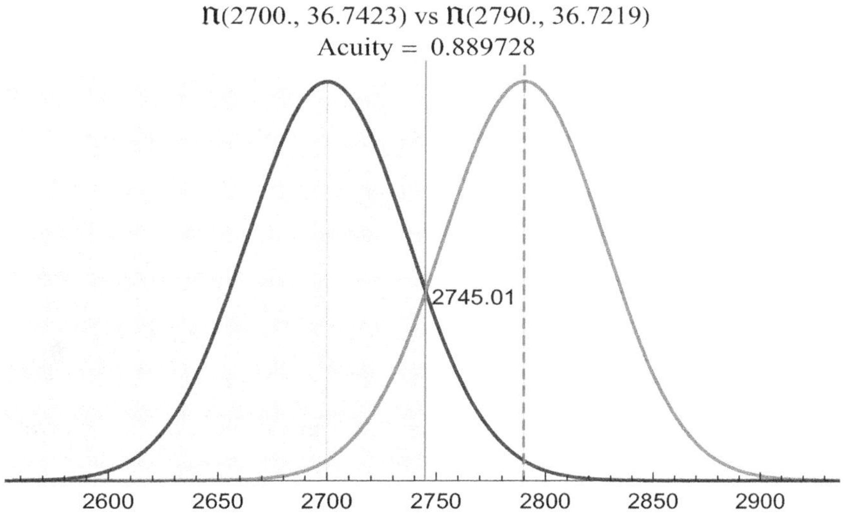

FIGURE 23.2. Negative Avoidance Precognition vs. Random

	Negative Avoidance Precognition The subject is able to avoid an aversive subliminal picture by making a choice before the computer makes a choice that will decide whether he is correct	
	$\boldsymbol{m > 2745}$	$\boldsymbol{m < 2745}$
$\mathfrak{h}_1$: Negative Avoidance Precognition is real	.88973	.11027
$\mathfrak{h}_2$: Negative Avoidance Precognition is just a statistical artifact	.11027	.88973
	say: "$\mathfrak{h}_1$ is true"	**say: "$\mathfrak{h}_2$ is true"**
	The Decision: Acuity = .88973	

TABLE 23.2. Negative Avoidance Precognition

The odds in favor correctly deciding in favor of 'Negative Avoidance" are 8.06852 to 1, far too low to register as a scientific finding.

There is a comparison that Bem did not make but inadvertently provided the data to test the a different hypothesis. Suppose that the computer, by some *paranormal* means could cause the subjects to slightly favor the "target" that the computer would choose.

$$\begin{aligned}\frac{828 + 1016 + 2790}{1560 + 2040 + 5400} &= \frac{4654}{9000} \\ &= .517\end{aligned}$$

The first numbers, in the numerator and denominator are the numbers from the trials involving erotic precognition, the second are those from negative and neutral trials from the same experiment, and the third, the trials from the negative avoidance experiment. The result is that over all the trials 51.7% were successes against 50% representing no precognition. The results are significant at the $\alpha = .0024$ level.

Figure and Table 23.3 show the result of choosing between the two using the optimum decision rule. The acuity is .921142.

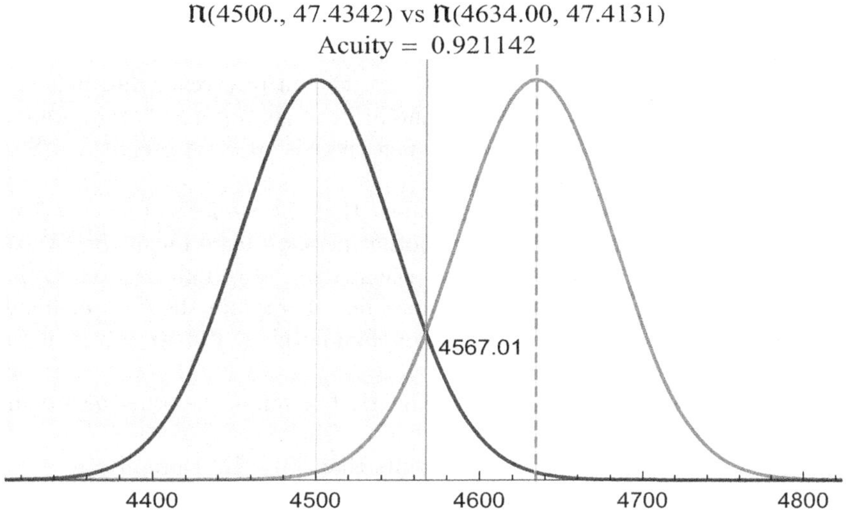

FIGURE 23.3. Computers have paranormal powers vs. Random

	Paranormal Computer Powers The Computer is able to influence the test subjects choices by paranormal means.	
	$m >$ **4567.01**	$m <$ **4567.01**
$\mathfrak{h}_1$: Computers have paranormal powers	.921142	.078858
$\mathfrak{h}_2$: The deviation from random is just a statistical artifact	.078858	.921142
	say: **"$\mathfrak{h}_1$ is true"**	**say:** **"$\mathfrak{h}_2$ is true"**
	The Decision: Acuity = .921142	

TABLE 23.3. Paranormal Computer Powers

Odds in favor of deciding correctly are only 11.68 to one, not in the big leagues, but better than the two similar ones published in the article.

The hypothesis, however, is surely as paranormal as it gets. Would Bem be influenced by the robust significance level? He has provided a rational—quantum entanglement [**4**, p. 422–423]. Quantum entanglement works in both directions. If it is a rational for Bem's results, it is a rational for this result.

Bem goes on to document a variety of other experiments all of which seemed to indicate something mysterious is going on. None of them pass muster as an individual finding. But he then combines the results by meta-analysis to come to what he considers a definitive conclusion. Would he be willing to include "Paranormal Computer Powers" in a meta-analysis? It is surely just as mysterious as anything included and individually has more empirical support.

The answer to meta-analysis is this: If a second experiment is identical to the first but for n, it could be treated as an addition to the first, the data from both can be pooled and analyzed together or the second could be treated a test of the first as shown in Chapter 16. If the second experiment is not essentially identical to the first, no further analysis of either type can be done. "Something mysterious is going on," does not meet the standard.

The episode shows three things. First, significance tests can trick a researcher into believing he's on to something when he isn't. Second, Bayesians can manipulate prior probabilities to get any result they want. Third, combining such diverse findings into a meta-analysis is suspect for the simple fact there is no way to replicate, "Something strange is going on." Individual findings can be replicated and should be!

In this is almost never done for two reasons. The first is fear that the model will not survive the test. The second is that in the social sciences there is no honor in replicating another person's experiment. See Makel, *et al.* [**32**].

When confronted with the thought of replication, social scientists tend to fall back on the argument that the model is not intended to predict the results of new data, but it's intended to *explain* the data at hand. This is an evasion. If the model is not probably true versus no model at all, it can explain nothing. If it were true, it can *explain* out-of-sample data.

Those who are not true believers in the paranormal take heart. Certain of the other results in Bem's paper have been replicated and failed, Ritchie, *et al.* [**40**].

It is not only fringe hypotheses that fail, mainstream results that often fail to replicate.

There has been an effort to encourage replications in social psychology.[1] Nosek and Lakens introduce a special issue of *Social Psychology*:

> This article introduces the first known journal issue in any discipline consisting exclusively of preregistered replication studies. It demonstrates that replications have substantial value, and that incentives can be changed.
>
> There are a number of advantages of performing direct replications, and publishing the results irrespective of the outcome. First, direct replications add data to increase precision of the effect size estimate via meta-analysis. Under some circumstances, this can lead to the identification of false positive research findings. Without direct replication, there is no way to confidently identify false positives.[**38**, p. 137]

One should be careful what one wishes for. By my count there are 22 hypotheses replicated in this journal issue. The original experiments were about as decisive as Bem's, that is, not decisive enough to be counted as definitive. The interpretations of the replications are not done correctly, as in Chapters 16 or 17, but rather based on significance testing.

On that basis, 6 of the 22 replicated successfully, but two of the replications were in theoretical conflict with each other. Fourteen simply did not replicate. Two did not replicate in original form, but did only after the experiment was slightly altered after the fact by excluding certain data.

Many of the original studies were the received knowledge of the field. One of the original papers had been cited over two thousand times. It is clear that the replicator wanted success, but 5 or 6 or 8 out of 22 is not a particularly encouraging result.

When successful replication did not occur, there was much hand waving by the researchers. In many instances, when there was a result slightly in the right direction, they would do a meta-analysis to find the true magnitude of the effect. In some cases, they suggested that more study was needed to explain the mediating factors that would explain the discrepancy between the original study and the replication.

There was never a statement that the original result was just a statistical fluke because faith in significance testing is the bedrock of the soft sciences. Just because the original studies were validated by significance testing on modest sized samples, there is little reason to believe they are true—significance testing promises a decisiveness that it cannot deliver.

The next chapter shows that hypotheses can be validated by significance testing with large sample size alone even if there is no there, there.

[1] There is also a program to replicate 50 cancer related studies, Errington, *et al.* [**13**].

CHAPTER 24

Meehl's Conjecture

Paul Meehl proposed the radical idea that, at least the field of psychology, the null hypothesis was almost always false and would be rejected by significance testing [**33**] [**34**].

His reasoning was that with a large enough sample size, even tiny differences would be shown to be significant and that tiny differences always exist. Given a large enough sample, he expected 100% of all possible two-tailed tests and 50% of one-tailed tests would reject the null hypothesis.

The assumption of significance testing is that the null hypothesis is almost always true and that only a small number of hypotheses would be rejected by significance testing, the proportion corresponding to the significance level.

Niels G. Waller, in "The fallacy of the null hypothesis in soft psychology" [**51**], put the question to a test. He chose 511 questions from the Minnesota Multiphasic Personality Inventory-2, chose the direction of the one-tailed test with $\alpha = .025$ at random, and tested for sex differences with the sample of 41,491 males and 39,994 females.

The results are shown in Figure 24.1. The graph shows normal approximations to two binomial distributions, the leftmost with $p = .025$ and the other with $p = .5$, both with $n = 511$. Because the approximations are truncated at zero, the usual practice of having a two-cut decision rule is replaced by a one-cut rule.

If Meehl were correct, one would expect 255.5 of the questions would show significant sex differences. If the tacit assumptions of significance testing were correct, one would expect only 12.775 of the questions to show significant sex differences. The cut comes at 70.531 or 13.8%. In fact, 46% of the questions showed significant sex differences, clearly in favor of Meehl's conjecture.

As a follow-up, Waller tested all possible pairs of questions, $567 \times 566 = 320,922$, for correlation or anti-correlation determined at random, using the 39,994 females in his database. Figure 24.2 shows the results of that test. If Meehl were correct, one would expect 160,460 of the question-pairs would show significant results. If the tacit assumptions of significance testing were correct, one would expect only 8023.05 of the questions to show significant results. The cut comes at 44,295.7 or 13.2%. In fact, approximately 47% of the question-pairs showed significant results, clearly in favor of Meehl's conjecture.

Figure 24.1 shows an acuity nearly equal to one $(1 - 5.887^{-59})$; the odds are in favor of Meehl's conjecture are 5.887×10^{59} to one. It is 2×10^{57} times stronger than the evidence in favor of General Relativity gleaned from the Eddington expedition and larger than ratio of the size of the universe to the size of a quark.

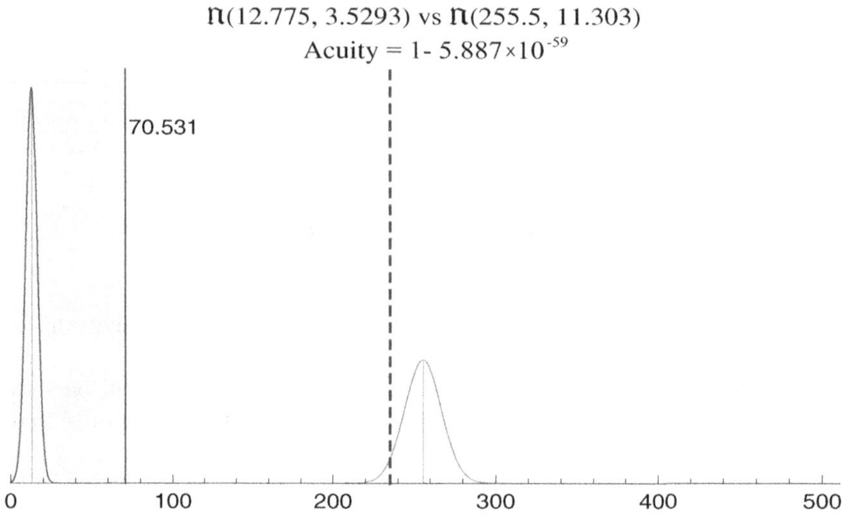

FIGURE 24.1. Normal Approximation to Meehl's Conjecture Using Sex Differences

	Meehl's Conjecture on Sex Differences Given large n, 50% of one-tailed t-tests applied to human data, will be significant.	
	$\boldsymbol{m < 70.531}$	$\boldsymbol{m > 70.531}$
$\mathfrak{h}_1$: One-tailed t-tests work the way they are supposed to.	$1 - 5.887 \times 10^{-59}$	5.887×10^{-59}
$\mathfrak{h}_2$: Meehl's Conjecture is correct.	5.887×10^{-59}	$1 - 5.887 \times 10^{-59}$
	say: **"$\mathfrak{h}_1$ is true"**	**say:** **"$\mathfrak{h}_2$ is true"**
	The Decision: Acuity $= 1 - 5.887 \times 10^{-59}$	

TABLE 24.1. Meehl's Conjecture on Sex Differences

Figure 24.2 also shows an acuity nearly equal to one ($> 1 - 10^{-200}$); the odds are in favor of Meehl's conjecture are are greater than 10^{200} to one, the about ratio of the size of the universe to the size of a quark raised to the fifth power, $10^{42} \times 10^{42} \times 10^{42} \times 10^{42} \times 10^{42}$ to one.

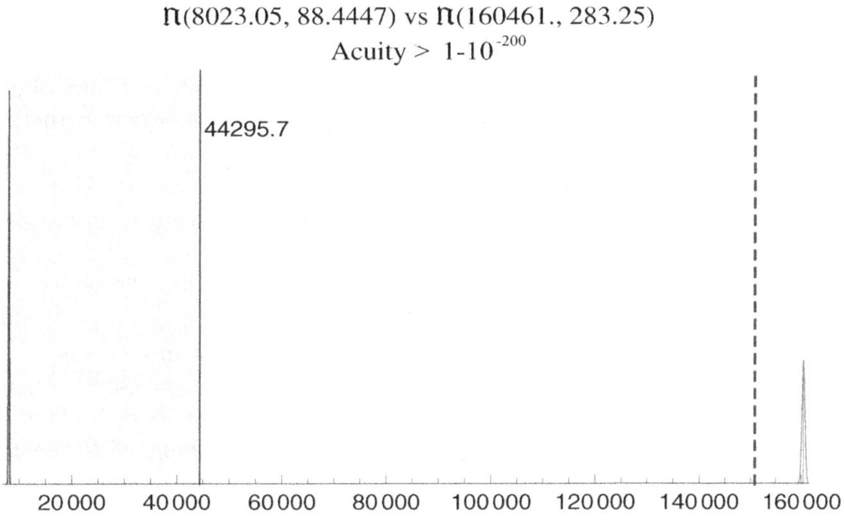

FIGURE 24.2. Normal Approximation to Meehl's Conjecture Using All Pairs of Questions

	Meehl's Conjecture on MMPI-2 Given large n, 50% of one-tailed t-tests applied to human data, will be significant.	
	$m < 44295.7$	$m > 44295.7$
$\mathfrak{h}_1$: One-tailed t-tests work the way they are supposed to.	$> 1 - 10^{-200}$	$< 10^{-200}$
$\mathfrak{h}_2$: Meehl's Conjecture is correct.	$< 10^{-200}$	$> 1 - 10^{-200}$
	say: "$\mathfrak{h}_1$ is true"	**say: "$\mathfrak{h}_2$ is true"**
	The Decision: Acuity $= > 1 - 10^{-200}$	

TABLE 24.2. Meehl's Conjecture on MMPI-2

Meehl restricted his conjecture the field of psychology, but it is important if applicable to a variety of disciplines. The most obvious is epidemiology, but in the study that examines a large number of subjects has something like 50% probability of finding statistically significant results.

In Chapter 21, we saw the same thing developed between intelligence and various social variables. With diligent work and large sample sizes, one can manufacture of philosophical superstructure on the basis of what is in effect the statistical equivalent of quantum foam.

In the current chapter, by calculating the probability that significance tests work as promised versus Meehl's conjecture, we have clear empirical evidence that significance testing just doesn't work the way it's intended to. In fact, significance testing is the greatest boon to academic publishing and the production of Ph.D.s that can be imagined.

Results should give one pause when evaluating large-scale studies focusing on factors that increase the likelihood of some disease or treatments that ameliorate a disease.

Think about how such studies progress. Data is collected from a large number of individuals over a long period of time. At some point a researcher with access to the data, decides to check to see whether there's a correlation between some data series that is been collected and some disease outcome.

The result should be suspect for two reasons:

- There is about a 50% chance that the answer to the question of significance will be yes irrespective of truth or theoretical import simply because of the large sample size.
- There is a vast gap between the significance level and the probability that the target hypothesis is true.

As shown in Chapter 16, if one had access to the raw data, one could analyze it in a slightly different way and get more accurate results. Unfortunately in the current climate, the data that would allow more acute evaluation is not presented.

If one could replicate the experiment with new data, one could put the hypothesis to a hard test and either pump the acuity up to an acceptable level or eliminate the head start implicit in data mining. That will never happen! The expense is too great simply by the fact that n is large and the time-frame is long. Additionally, there is no honor in exactly replicating another's study.

The result is that another feeble unchallenged hypothesis is added to the list of received knowledge.

Part V

Less Than Meets the Eye

CHAPTER 25

Introduction to Part V

There has always been a desire to make probability statements about the truth of hypotheses. One strategy has increasingly gained adherents—Bayesian Theory. It is very seductive because it is well adapted to ordering one's own beliefs. It's scope has been expanded to claim that it can provide interpersonal consensus and that it can produce valid probability statements about the state of the world.

Neither claim is true.

CHAPTER 26

Bayesian Theory

Bayes' Theorem vs. Bayesian Theory

One must distinguish between Bayes' theorem and Bayesian Theory. Bayes' theorem is simply a true theorem derivable from the axioms of probability. Bayesian Theory is based on the willingness to interpret Bayes' theorem in a certain way.

The probability axioms have many manifestations. To understand the failings of Bayesian Theory, one must draw a distinction between two of them. It will be useful to draw a distinction in nomenclature between physical and Bayesian probabilities.

Physical Probability vs. Bayesian Probability
$\mathrm{P}_{\mathfrak{h}_i}(e)$ vs. $\mathbb{P}_{\widehat{\mathfrak{b}}}(\mathfrak{h}_i)$ & $\mathbb{P}_{\widehat{\mathfrak{b}}}(\mathfrak{h}_i \mid e)$

$\mathrm{P}_{\mathfrak{h}_i}(e)$ is the probability of e in a probability system that acts on physical world as delineated in hypothesis, $\mathfrak{h}_i$. $\mathrm{P}(\mathfrak{h}_i)$ does not exist; $\mathfrak{h}_i$ is not an element of the probability space. $\mathfrak{h}_i$ is either true or false.

Physical probability acts on the world in a way similar to Newton's gravitation. It is a "force" that acts on individual instances in a way that if $\mathfrak{h}_i$ is true, for hypothetical infinite sequences and all infinite subsequences that can be specified *ex ante*, as $n \to \infty$, the relative frequency of event, e, will tend as a limit to $\mathrm{P}_{\mathfrak{h}_i}(e)$.

$\mathbb{P}_{\widehat{\mathfrak{b}}}(\mathfrak{h}_i \mid e)$ is a probability system based on the beliefs of person, $\widehat{\mathfrak{b}}$, who believes each $\mathfrak{h}_i$ to be true, *a priori*, with subjective probability, $1 > \mathbb{P}_{\widehat{\mathfrak{b}}}(\mathfrak{h}_i) \geq 0$ and uses Bayes' theorem to modify initial beliefs given the occurrence of e. Hypotheses are enclosed in the same mantle as physical events.

Both $\mathrm{P}_{\mathfrak{h}_i}$ and $\mathbb{P}_{\widehat{\mathfrak{b}}}$ follow the probability axioms, but have different meanings.

There have been objections to physical probability on the grounds that it is hard to *exactly* define it. I follow Wittgenstein: The meaning is how it is used.

When teased out into a chain of reasoning, the Bayesian case has three main propositions and one auxiliary proposition, shown in **boldface**.[1]

(1) **When people talk about the probability, no matter what they claim, the are talking about a subjective degree of belief.**[2 3]

(2) There is only *one* kind of probability—a subjective feeling reflecting personal belief manifested either by stating "I believe the probability that hypothesis, $\mathfrak{h}$ is true or event, e will occur, to such and such a degree" or by a willingness to bet on the truth of the hypothesis or occurrence of the event, at particular odds. All definitions of probability, in the end, reduce to this ***subjective*** uncertainty, so it is permitted to put intrinsically dissimilar things inside the same probability space.

(3) **The Bayesian believes that it makes sense to assign a real number $1 > \mathbb{P}(\mathfrak{h}) \geq 0$ to each hypothesis, $\mathfrak{h} \in \mathfrak{H}$, such that**

$$\sum_{\mathfrak{h}\in\mathfrak{H}} \mathbb{P}(\mathfrak{h}) = 1 \textbf{ or } \int_{\mathfrak{h}\in\mathfrak{H}} \mathbb{P}(\mathfrak{h})d\mathfrak{h} = 1,$$

prior to any evidence.[4]

(4) When all else fails, imagined betting behavior is a way to tease out subjective degree of belief.[5]
   - Betting behavior must be consistent with the probability calculus to avoid a "Dutch book" bet.[6 7]
   - If A and B are mutually exclusive and a bet on A plus a bet on B is equivalent to a bet on (A or B), betting behavior must be consistent with the probability calculus.[8]

---

[1] Any particular account may omit some of these points or contain more or insist on the primacy of one or another. To my mind the best exposition is in Howson and Urbach [**21**, chapters 3, 4, 10, and 11].

[2] Historically, probability theory was developed to rationalize betting on "games of chance" such as cards or dice. It was intended to be a mathematical reflection of reality that would facilitate winning bets. Bayesians have made the reflection in the mirror into reality.

[3] The American Psychological Association has recently given qualified endorsement to Bayesian procedures, however in [**26**, pages 291–292] Kline does not seem to know the meaning of subjective and the act of guessing.

> "A longtime objection is that Bayesian methods are associated with the subjectivist view of probability. Recall that the subjectivist view does not distinguish between repeatable and unrepeatable (unique) events, and probabilities are considered as degree of personal belief that may vary from person to person. There is a perception among those unfamiliar with Bayesian statistics that prior probabilities of hypotheses are wholly subjective guesses just plucked out of the thin air, perhaps to suit some whim or prejudice.
>
> "These perceptions of Bayesian statistics are false. If nothing is known about the hypotheses, the researcher has little choice other than to guess about plausibility. But it is rare for researchers to have absolutely no previous information on which to base estimates of prior probabilities."

[4] Evidence has a special meaning in this situation. It is some event that has a likelihood of occurring given the truth of each of the theories in question. This is how the theories come to have links to the real world. I say "prior to *any* evidence" because, if a specific bit of evidence is included in the prior probabilities, Bayes' theorem, running in reverse, can be used to remove it, $\mathbb{P}(\mathfrak{h}) = \frac{\mathbb{P}(\mathfrak{h}|e)\mathbb{P}(e)}{\mathbb{P}(e|\mathfrak{h})}$. At some point, when all such evidence has been removed, the Bayesian is faced with assigning probability based on something other than their links to the real world.

[5] Cf. Lindley [**31**, pages 33–36] or Howson and Urbach [**21**, pages 56–59] or De Finetti [**7**, pages 3–7].

[6] A Dutch book bet is a combination of bets that, if accepted together, one cannot win. An example would be simultaneously accepting a bet that a coin (perhaps not a fair coin) would come heads at 2:1 odds and that it would come up tails, also at 2:1 odds. However, Kahneman and Tversky [**47**] [**48**] [**25**] [**49**] have shown that real people don't estimate probability or bet in a way consistent with the probability axioms.

[7] Milne [**35**] and Hájek [**19**] have posed the problem of avoiding a Dutch book bet as the problem of outwitting an omniscient bookie. This is a unnecessary; deciding on one's betting odds is intrinsically a solitary endeavor. Having examined the proofs that one's betting odds should be consistent with the probability axioms, one would, by interior dialogue, decide the appropriate odds for particular bet. A bet proposed by a bookie, omniscient or not, good, bad, or neutral, would be accepted or *not*. Self consistency is the only criterion.

[8] See Skyrms [**44**, chapter VI]. The argument loses some force because it assumes that the number we are trying to *prove* is a probability, is very close to being a probability already. This is because he calculates the expected value of bets based *epistemic plausibility/probability*

- Thus betting behavior and beliefs about uncertainty, must follow the axioms of probability in order to avoid inconsistencies.

(5) There is no need to posit any other kind of probability but the subjective because:
- Subjective probability exists where no "objective" probability is thought to exist.[9] Thus "subjective" probability must exist and be logically prior to the "objective."
- Whenever "objective" probability is thought to exist "subjective" probability can be posited, numerically equal to the "objective" because, if the "subjective" is not equal to the "objective," betting losses occur. Thus "objective" probability is unnecessary.
- In situations where some kind of *objective* probability is posited, in reality, there is only the *subjective*. When we assert that a coin has the propensity to come up "heads" half the time, we are unjustifiably objectifying our our subjective beliefs.
- Subjective probability is the only way to countenance probability of a single occurrence.

(6) Logically pushed to the extreme, because there is only *one* kind of probabilistic uncertainty, everything uncertain for person, $\widehat{\mathfrak{b}}$, is can be made part of one universal probability distribution, $\mathbb{P}_{\widehat{\mathfrak{b}}}(\cdot \mid \cdot)$ All things, no matter how dissimilar, can coexist within $\mathbb{P}_{\widehat{\mathfrak{b}}}(\cdot \mid \cdot)$, if $\widehat{\mathfrak{b}}$ believes it to be so.
- This means that the probability that $\widehat{\mathfrak{b}}$ will draw a full house to a pair in poker can be part of the same probability scheme as the probability that a theorem is true given that mathematicians have tried to prove it for several hundred years, without success, and have also failed to find a counter-example for just as long;
- or that George Washington would have been able to throw a cherry pie across the Delaware given that he was able to throw a dollar across the Potomac;
- or that Einstein's theory of general relativity is true, even though $\widehat{\mathfrak{b}}$ does not imagine that its truth is determined by anything like the roll of a die or draw of a card.
- It all depends on which part of the probability space $\widehat{\mathfrak{b}}$ chooses to attend to at the time.[10]
- One doesn't even use Bayes' theorem to revise probabilities. One can just change his mind as long he changes it all around to avoid a "Dutch Book" bet.

(7) **The incorporation of evidence is seen as the modification of the initial probabilities by Bayes' theorem and that process inevitably leads to convergence to the correct posterior probabilities:**

(a) New evidence modifies the probabilities such that the probability of hypothesis, $\mathfrak{h}$, after evidence, e, becomes known, is $\mathfrak{h}$ conditioned on e, $\mathbb{P}(\mathfrak{h} \mid e)$, otherwise a "Dutch Book" bets can be proposed between the probability of $\mathfrak{h}$ before and after evidence, e, is collected, Skryms [**44**, Chapter VI.7.].

(b) As evidence accumulates, the posterior probability, $\mathbb{P}(\mathfrak{h} \mid e)$ approaches certainty. Because

$$\mathbb{P}(\mathfrak{h} \mid e) = \frac{\mathbb{P}(\mathfrak{h}) \times \mathbb{P}(e|\mathfrak{h})}{(\mathbb{P}(\mathfrak{h}) \times \mathbb{P}(e|\mathfrak{h})) + (\mathbb{P}(\neg\mathfrak{h}) \times \mathbb{P}(e|\neg\mathfrak{h}))}$$

Thus, if $\mathfrak{h}$ is true and $\mathbb{P}(\mathfrak{h}) \neq 0$, as e accumulates, $\mathbb{P}(e \mid \mathfrak{h})$ will outstrip $\mathbb{P}(e \mid \neg\mathfrak{h})$ and thus $\mathbb{P}(\mathfrak{h} \mid e) \to 1$. On the basis of this the Bayesian would claim that, so long as they are not one or zero, prior probabilities make little difference because sufficient collection of data will overwhelm them.

(c) Thus Bayesian Theory is based on the belief that it makes sense to say that $\mathbb{P}(\mathfrak{h} \mid e) = \frac{\mathbb{P}(\mathfrak{h})\mathbb{P}(e|\mathfrak{h})}{\mathbb{P}(e)}$. This is a true consequence of Bayes' theorem, but Bayes' theorem is based on a single clearly defined probability, whereas Bayesian Theory depends on an odd hybrid:

---

just as if it *were* a probability in the physical sense. The concept and calculation of expected value and fairness presupposes something like it. The justification for using *epistemic probability* for placing bets is unclear. Even granting the logic of this move, a bet could seem fair based on *epistemic probability*, but it will only *be* fair if the numbers are correct.

[9]For example, in the case of scientific hypotheses, it is not *really* believed that the true hypothesis is determined by a random event at the moment before evidence has begun to be collected so the prior probabilities must have another source.

[10]If one wished to insulate a from the influence of b, one has only to set the conditional probability in this way: $\mathbb{P}(a \mid b) = \mathbb{P}(a)$ and $\mathbb{P}(b \mid a) = \mathbb{P}(b)$.

- where $\mathbb{P}(\mathfrak{h})$ is the *subjective* probability of the truth of an hypothesis;
- where $\mathbb{P}(e \mid \mathfrak{h})$ is a *subjective* probability of an *objective* event, the value determined by the *meaning* of $\mathfrak{h}$, which deals with physical probability in the sense of propensity or relative frequency;
- and where $\mathbb{P}(\mathfrak{h} \mid e)$ and $\mathbb{P}(e)$ are a combination of the *subjective* probability and *objective* probability.[11]

(8) **The auxiliary proposition is this: If we assume that science is following the Bayesian paradigm, many of the problems in the philosophy and history of science are explained.**[12]

If I say that, to me, partial knowledge or willingness to bet or degree of belief is not the same thing as what goes on in games of chance; a Bayesian will say that I *ought* to believe they are the same because, if I do, I can:

(a) explain how scientists opinions converge to a consensus, presumably correct,
(b) explain why, if a theory successfully predicts an unlikely outcome it is strongly confirmed,
(c) explain why some evidence is of little relevance, *i.e.* the existence of a white rabbit does little to confirm the proposition, "All ravens are black,"
(d) and explain why evidence can confirm or disconfirm one part of a larger theory more than others, why some modifications of previous theories are acceptable and some are viewed as unacceptably *ad hoc*.

The difficulty is that the three main propositions are false and the auxiliary proposition does not deliver what it promises.

Section 26.1 addresses the assertion that subjective probability is a privileged interpretation of the probability axioms.

Section 26.2 shows that the claims of convergence are overstated.

Section 26.3 puts Rev. Bayes in the position of ferreting out the rules of the universe at the moment of their creation.

26.1. Subjective Probability *Not* Privileged

In this section we will examine the idea that subjective probability is a valid substitute for those other things that people might mean when they use the term probability.

In the first subsection, we take a look at the Bayesian assertion that vague initial uncertainty can be adequately represented by probabilities based on the assignment of prior probabilities. We will find the ordinary form of uncertainty already has an expression in the probability calculus and that it cannot be represented by assigning prior probabilities to hypotheses in a way that can use Bayes' theorem to accommodate the accretion of evidence.

In the second subsection, we will construct an alternative probability that could never be mistaken for subjective degree of belief or a physical propensity, but still has all the syntactic attributes that recommend the probability calculus as the only correct form for representing subjective degree of belief. Then we will show that it is all too easy to splice such probability system to another probability system that has a different semantic content. The resultant system will be syntactically correct, but semantically meaningless. In other words, in the resultant system, $\mathrm{P}\,(\textit{whatever})$ is just a number.

Uncertainty Without Priors. The Bayesian bases initial probabilities on the weight of prior knowledge or on an assignment of probabilities designed to represent the absence of any prior knowledge. If there are an infinite number hypotheses to be considered, there is no unique way to assign initial probabilities. If there are finite choices, the standard procedure is to assign probability equally to all, but this does not really represent total absence of knowledge and what does, does not allow transforming likelihood ratios into posterior probabilities.

Consider this:

[11] In our terminology:

$$\mathbb{P}(\mathfrak{h} \mid e) = \frac{\mathbb{P}(\mathfrak{h})\,\mathbb{P}(e|\mathfrak{h})}{\mathbb{P}(e)} \stackrel{?}{=} \frac{\mathbb{P}(\mathfrak{h}) \times \mathrm{P}_{\mathfrak{h}}(e)}{(\mathbb{P}(\mathfrak{h}) \times \mathrm{P}_{\mathfrak{h}}(e)) + (\mathbb{P}(\neg\mathfrak{h}) \times \mathrm{P}_{\neg\mathfrak{h}}(e))}$$

[12] A good brief summary is available in Howson and Urbach [**21**, chapter 4]

EXAMPLE 26.1.

Imagine $\widehat{\mathfrak{b}}$, a Bayesian, *who believes that there are two types of urns*: *Green* ones with probabilities $\frac{1}{3}$, $\frac{1}{3}$, $\frac{1}{3}$ to produce *Black*, *Grey*, or *White* balls and *Red* ones with $\frac{1}{4}$, $\frac{1}{2}$, $\frac{1}{4}$ probability to produce *Black*, *Grey*, or *White* balls. He further believes that the two types of urns are *equally likely to occur.*

Imagine $\widehat{\mathfrak{o}}$, someone who believes in a more objective view of probability, at least is not a Bayesian. $\widehat{\mathfrak{o}}$ believes *all urns are the same type*, agrees with $\widehat{\mathfrak{b}}$ as to contents of *Green* and *Red*, but is *undecided* as to whether all are *Green* or *Red*.

Imagine this dialogue:

$\widehat{\mathfrak{b}}$: What you need to do is to imagine yourself betting on whether *Red* or *Green* is true. Once know your betting odds, you can calculate your subjective probabilities and we can get down to work, drawing balls. This is what I believe:

$$\mathbb{P}_{\widehat{\mathfrak{b}}}(Green) = \mathbb{P}_{\widehat{\mathfrak{b}}}(Red) = \frac{1}{2}$$

$$\mathbb{P}_{\widehat{\mathfrak{b}}}(Red\ Green) = 0$$

$$\mathbb{P}_{\widehat{\mathfrak{b}}}(Black \mid Green) = \mathbb{P}_{\widehat{\mathfrak{b}}}(Grey \mid Green) = \mathbb{P}_{\widehat{\mathfrak{b}}}(White \mid Green) = \frac{1}{3}$$

$$\mathbb{P}_{\widehat{\mathfrak{b}}}(Black \mid Red) = \mathbb{P}_{\widehat{\mathfrak{b}}}(White \mid Red) = \frac{1}{4}, \quad \mathbb{P}_{\widehat{\mathfrak{b}}}(Grey \mid Red) = \frac{1}{2}$$

Your problem is that you have not plumbed your own mind deeply enough. A little extra work is necessary to form your belief, that *Red* is true or *Green* is true, into a probability calculus.

$\widehat{\mathfrak{o}}$: I don't understand what you mean. My beliefs are a logical disjunction that does not require introspection to translate it into a probability calculus; *logical disjunction already is a statement in the probability calculus* (Theorem A.90, page 187). Specifically:

There are two elements in the probability space of urns, *Green* and *Red*.

$$\mathbb{P}_{\widehat{\mathfrak{o}}}(Green \mid Red) = \mathbb{P}_{\widehat{\mathfrak{o}}}(Red \mid Green) = 0$$

$$\mathbb{P}_{\widehat{\mathfrak{o}}}(Red\ Green) = 0$$

$$\mathbb{P}_{\widehat{\mathfrak{o}}}(Red) + \mathrm{P}_{\mathfrak{o}}(Green) = 1$$

$$\mathbb{P}_{\widehat{\mathfrak{o}}}(\neg(\neg Red\ \neg Green)) = 1$$

$$\mathbb{P}_{\widehat{\mathfrak{o}}}(Green) = 1 \text{ or } \mathbb{P}_{\widehat{\mathfrak{o}}}(Red) = 1$$

The first proposition indicates that if the urn is one color there is zero probability that is the other color. The second proposition indicates that it is not the case that the urn is both colors. The third proposition indicates that the sum of the probabilities are 1. The fourth proposition indicates that the probability space consists of *Red* or *Green* and nothing else. The fifth, that either *Green* or *Red* is true. It is a very small probability space.

Green and *Red* are defined as:

$$\mathrm{P}_{Green}(Black) = \mathrm{P}_{Green}(Grey) = \mathrm{P}_{Green}(White) = \frac{1}{3}$$

and

$$\mathrm{P}_{Red}(Black) = \mathrm{P}_{Red}(White) = \frac{1}{4} \text{ and } \mathrm{P}_{Red}(Grey) = \frac{1}{2}$$

The assignment of specific probabilities to *Green* and *Red*, even though they are equal, assumes extra knowledge, just as would $\mathbb{P}_{\widehat{\mathfrak{o}}}(Green) = .99$ and $\mathbb{P}_{\widehat{\mathfrak{o}}}(Red) = .01$.

$\widehat{\mathfrak{b}}$: But if you don't assign individual probabilities to *Red* and *Green*, you can't use Bayes' theorem to process evidence and you will never be able to decide whether *Red* or *Green* is more probable after the collection of evidence.

$\widehat{\mathfrak{o}}$: The reason I don't assign probabilities to *Red* and *Green* is because I don't believe they can be specified *a priori*. About the processing of evidence, there is no problem. Chapter 6 shows how one can make probability statements about hypotheses without bothering to make anything up!

Why should I make up a number that doesn't correspond to my uncertainty just so I can make a calculation that I don't need. The results would just be meaningless numbers.

$\widehat{\mathfrak{b}}$: If you did so, we could collect evidence together and our opinions would converge.

$\widehat{\mathfrak{o}}$: That's not so clear either. If you mean that the accretion of evidence decides empirical questions, none would disagree.

However, you seem to be making a claim of numerical correctness. If $\mathfrak{h}$ is an event, that is the true hypothesis is picked at random as if it were a game of chance, $\mathbb{P}_{\widehat{\mathfrak{b}}}(\mathfrak{h} \mid e)$ is correct only if $\mathbb{P}_{\widehat{\mathfrak{b}}}(\mathfrak{h})$ is correct. If that is not the case, it is hard to see why $\mathbb{P}_{\widehat{\mathfrak{b}}}(\mathfrak{h} \mid e)$ is not just an arbitrary number. Positing some $\mathbb{P}_{\widehat{\mathfrak{b}}}(\mathfrak{h})$ gives a rationale for normalizing $\mathbb{P}_{\widehat{\mathfrak{b}}}(e \mid \mathfrak{h})$ so that you can say that the posterior probability is thus and so, which is impossible, using Bayes' Theorem, without the fiction of prior probabilities.

I grant that the Bayesian exercise could be of use in keeping your thoughts in order and avoiding Dutch Book bets, but the addition of $\mathbb{P}_{\widehat{\mathfrak{b}}}(\mathfrak{h})$ adds nothing to the discovery of real world truths not provided by e.

Using the method explicated in Chapter 6, our opinions after viewing evidence would be identical; the only convergence would be convergence toward certainty.

If by convergence, you meant that if we started with different prior probabilities, our posterior probabilities would converge Section 26.2 below discredits that notion.

There is simply no justification for insisting that an hypothesis that already can be expressed in the probability calculus *must* be replaced by another to which it is not equivalent. A disjunction of hypotheses is directly translatable into the probability calculus. It is a kind of uncertainty that is inconsistent with the existence of the initial probabilities of the Bayesian sort.

Semantics Trumps Syntax. The point of this subsection is to show that one can construct a probability system completely without reference to relative frequency or propensity or subjective degree of belief.

EXAMPLE 26.2.

Suppose that there are *Red* urns and *Green* urns. Each urn contains balls that are identical except that some are colored *Silver* and some are colored *Gold*. The urns and balls are labeled with their color in English.

The probability system to be demonstrated is based on the *colors* of the urns and the balls, specifically how the names of the colors are spelled in a the English language. This probability system is of no use except to show that it *can* be constructed, could *never* be mistaken for any of the usual interpretations of the probability calculus, and that if one of the pertinent parameters, color of the urns or color of the balls, were changed from the current example, we would know how to calculate a revised scheme. For example, if the *Red* urn were *Blue* instead, we would know how to calculate the new probability system.

Because it follows the probability calculus, it has all the syntactic properties that the Bayesian believes delimit subjective degree of belief, he might be tempted to wrap it in the cloak of subjective probability.

Our example is so egregious that no one would dare to attempt this maneuver.

Let this color-based system be $P_9(\cdot \mid \cdot)$.

In order to get to $P_9(\cdot)$ we take a detour through $\text{modulus}_9(\cdot)$. First we deal with positive integers: The $\text{modulus}_9(\textit{positive integer})$ is the result of repeatedly subtracting nine from the number until remainder is less than ten.

Next we deal with the alphabet. The $\text{modulus}_9(\textit{letter})$ is derived from this table:

Mapping of Alphabet to $\textbf{modulus}_9(\textit{letter})$

1	2	3	4	5	6	7	8	9
A	B	C	D	E	F	G	H	I
J	K	L	M	N	O	P	Q	R
S	T	U	V	W	X	Y	Z	

Next words: The $\text{modulus}_9(\textit{word})$ is the sum of the $\text{modulus}_9(\textit{letters in the word})$ from which nine is repeatedly subtracted until the remainder is less than ten.

For our example:

$$\text{modulus}_9(\textit{Red}) = 9 + 5 + 4 = 9$$

$$\text{modulus}_9(\textit{Green}) = 7 + 9 + 5 + 5 + 5 = 31 = 4$$

$$P_9(\textit{urn}) \text{ is proportional to } \text{modulus}_9(\textit{urn})$$

$$P_9(\textit{Red}) = \frac{9}{13}$$

$$P_9(\textit{Green}) = \frac{4}{13}$$

The $\text{modulus}_9(\textit{ball})$ is the $\text{modulus}_9(\textit{label of the ball})$. In our example:

$$\text{modulus}_9(\textit{Silver}) = 1 + 9 + 3 + 4 + 9 = 26 = 8$$

$$\text{modulus}_9(\textit{Gold}) = 7 + 6 + 3 + 4 = 20 = 2$$

The modulus of a ball contained in a particular colored urn, $\text{modulus}_9(\textit{ball} \mid \textit{urn})$ is:

$$\frac{\text{modulus}_9(\text{modulus}_9(\textit{ball}) + \text{modulus}_9(\textit{urn}))}{\text{modulus}_9(\textit{urn})}.$$

The $P_9(\textit{ball} \mid \textit{urn})$ is proportional to $\text{modulus}_9(\textit{ball} \mid \textit{urn})$. For example:

$$\begin{aligned}\text{modulus}_9(\textit{Silver} \mid \textit{Red}) &= \frac{\text{modulus}_9(\text{modulus}_9(\textit{Silver}) + \text{modulus}_9(\textit{Red}))}{\text{modulus}_9(\textit{Red})} \\ &= \frac{\text{modulus}_9(8+9)}{9} = \frac{8}{9}\end{aligned}$$

$$\begin{aligned}\text{modulus}_9(\textit{Gold} \mid \textit{Red}) &= \frac{\text{modulus}_9(\text{modulus}_9(\textit{Gold}) + \text{modulus}_9(\textit{Red}))}{\text{modulus}_9(\textit{Red})} \\ &= \frac{\text{modulus}_9(2+9)}{9} = \frac{2}{9}\end{aligned}$$

$$\begin{aligned}
\mathrm{P}_9(Silver \mid Red) &= \frac{\mathrm{modulus}_9(Silver \mid Red)}{\mathrm{modulus}_9(Silver \mid Red) + \mathrm{modulus}_9(Gold \mid Red)} \\
&= \frac{\frac{8}{9}}{\frac{8}{9} + \frac{2}{9}} = \frac{4}{5}
\end{aligned}$$

$$\begin{aligned}
\mathrm{P}_9(Gold \mid Red) &= \frac{\mathrm{modulus}_9(Gold \mid Red)}{\mathrm{modulus}_9(Silver \mid Red) + \mathrm{modulus}_9(Gold \mid Red)} \\
&= \frac{\frac{2}{9}}{\frac{8}{9} + \frac{2}{9}} = \frac{1}{5}
\end{aligned}$$

$$\begin{aligned}
\mathrm{modulus}_9(Silver \mid Green) &= \frac{\mathrm{modulus}_9(\mathrm{modulus}_9(Silver) + \mathrm{modulus}_9(Green))}{\mathrm{modulus}_9(Green)} \\
&= \frac{\mathrm{modulus}_9(8+4)}{4} = \frac{3}{4}
\end{aligned}$$

$$\begin{aligned}
\mathrm{modulus}_9(Gold \mid Green) &= \frac{\mathrm{modulus}_9(\mathrm{modulus}_9(Gold) + \mathrm{modulus}_9(Green))}{\mathrm{modulus}_9(Green)} \\
&= \frac{\mathrm{modulus}_9(2+4)}{4} = \frac{3}{2}
\end{aligned}$$

$$\begin{aligned}
\mathrm{P}_9(Silver \mid Green) &= \frac{\mathrm{modulus}_9(Silver \mid Green)}{\mathrm{modulus}_9(Silver \mid Green) + \mathrm{modulus}_9(Gold \mid Green)} \\
&= \frac{\frac{3}{4}}{\frac{3}{4} + \frac{3}{2}} = \frac{1}{3}
\end{aligned}$$

$$\begin{aligned}
\mathrm{P}_9(Gold \mid Green) &= \frac{\mathrm{modulus}_9(Gold \mid Green)}{\mathrm{modulus}_9(Silver \mid Green) + \mathrm{modulus}_9(Gold \mid Green)} \\
&= \frac{\frac{3}{2}}{\frac{3}{4} + \frac{3}{2}} = \frac{2}{3}
\end{aligned}$$

To summarize $\mathrm{P}_9(\cdot \mid \cdot)$:

$$\mathrm{P}_9(Red) = \frac{9}{13} \qquad \mathrm{P}_9(Green) = \frac{4}{13}$$

$$\mathrm{P}_9(Silver \mid Red) = \frac{4}{5} \qquad \mathrm{P}_9(Silver \mid Green) = \frac{1}{3}$$

$$\mathrm{P}_9(Gold \mid Red) = \frac{1}{5} \qquad \mathrm{P}_9(Gold \mid Green) = \frac{2}{3}$$

Extending the results by means of the probability axioms:

$$\mathrm{P}_9(Silver) = \frac{128}{195} \qquad \mathrm{P}_9(Gold) = \frac{67}{195}$$

$$\mathrm{P}_9(Red \mid Silver) = \frac{27}{32} \qquad \mathrm{P}_9(Green \mid Silver) = \frac{5}{32}$$

$$\mathrm{P}_9(Red \mid Gold) = \frac{27}{67} \qquad \mathrm{P}_9(Green \mid Gold) = \frac{40}{67}$$

For other combinations:

$$\mathrm{P}_9(\cdot \mid \cdot) = 0 \text{ or } 1$$

The mention of urns and balls is completely gratuitous; the probabilities are built on words. There is no intention that a ball be drawn from an urn. Now this has been a long excursion, seemingly for little result. We have constructed an alternative probability, $\mathrm{P}_9(\cdot \mid \cdot)$ that doesn't seem to have any importance. This is not the same thing as saying that it has no meaning. If someone made an error in calculating this strange probability, we would be able to tell immediately. We can even create related probability distributions in different foreign languages.

What we could *not* do is mistake $\mathrm{P}_9(\cdot \mid \cdot)$ as saying anything about the long run behavior of relative frequencies, propensities, or subjective beliefs.

Similarly, even though it can be used for making bets, maintaining that $\mathrm{P}_9(\cdot \mid \cdot)$ was a reasonable subjective degree of belief would probably be an express ticket to the insane asylum, but not on syntactic grounds. Whatever it means, subjective degree of belief, does not depend on color names in the English language.

It is clear that $\mathrm{P}_9(\cdot \mid \cdot)$ is not equivalent to physical probability and is not equivalent to subjective degree of belief, in the usual sense. It is just an arbitrary mapping of strings of symbols into non-negative numbers, whose sum is one.

EXAMPLE 26.3.

Suppose that there is one *Red* urn and is one *Green* urn. Each urn contains infinitely many balls that are identical except that some are colored *Silver* and some are colored *Gold.*

In the *Red* urn, the balls are in the ratio of four silver balls for every gold ball.

In the *Green* urn, the balls are in the ratio of two gold balls for every silver ball.

If the urn is *Red*, the probability of drawing a ball of a particular color is this:

$$\mathrm{P}_{Red}(Silver) = \frac{4}{5}$$
$$\mathrm{P}_{Red}(Gold) = \frac{1}{5}$$

If the urn is *Green*, probability of drawing a ball of a particular color is this:

$$\mathrm{P}_{Green}(Silver) = \frac{1}{3}$$
$$\mathrm{P}_{Green}(Gold) = \frac{2}{3}$$

There is a straight forward assignment of probabilities to these events based on the hypothesized relative frequency of their occurrence. Balls are drawn from the urns in such a way that successive draws are independent. These two probability systems are extended to multiple draws in the usual way so that the probability is distributed equally among possible permutations of *Silver* and *Gold.*

If i *Silver* balls and j *Gold* balls are drawn from the *Red* urn, their joint probability is:

$$\Big(\mathrm{P}_{Red}(Silver)\Big)^i \times \Big(\mathrm{P}_{Red}(Gold)\Big)^j = \left(\frac{4}{5}\right)^i \left(\frac{1}{5}\right)^j$$

If i *Silver* balls and j *Gold* balls are drawn from the *Green* urn, their joint probability is:

$$\Big(\mathrm{P}_{Green}(Silver)\Big)^i \times \Big(\mathrm{P}_{Green}(Gold)\Big)^j = \left(\frac{1}{3}\right)^i \left(\frac{2}{3}\right)^j$$

These two probability systems are just the kind that was the starting point of probability theory. However, it is possible, even if probability systems have totally different meanings, to splice them together into a hybrid probability system. That is what we will do now.

We know from Theorem A.127, page 193 that we can join these two ordinary probability systems, $\mathrm{P}_{Red}(\cdot)$ and $\mathrm{P}_{Green}(\cdot)$, with part of $\mathrm{P}_9(\cdot \mid \cdot)$, to form a *composite* probability system, $\mathrm{P}_{inepta}(\cdot|\cdot)$:

$$\mathrm{P}_{inepta}(Silver \mid Red) = \mathrm{P}_9(Red)\,\mathrm{P}_{Red}(Silver) = \frac{9}{13} \times \frac{4}{5} = \frac{36}{65}$$
$$\mathrm{P}_{inepta}(Gold \mid Red) = \mathrm{P}_9(Red)\,\mathrm{P}_{Red}(Gold) = \frac{9}{13} \times \frac{1}{5} = \frac{9}{65}$$
$$\mathrm{P}_{inepta}(Silver \mid Green) = \mathrm{P}_9(Green)\,\mathrm{P}_{Green}(Silver) = \frac{4}{13} \times \frac{1}{3} = \frac{4}{39}$$
$$\mathrm{P}_{inepta}(Gold \mid Green) = \mathrm{P}_9(Green)\,\mathrm{P}_{Green}(Gold) = \frac{4}{13} \times \frac{2}{3} = \frac{8}{39}$$

If $\mathrm{P}_9(\cdot \mid \cdot)$ is absurd, the composite, $\mathrm{P}_{inepta}(\cdot \mid \cdot)$ is *complete* nonsense. But both are equally consistent with the probability axioms.

$\mathrm{P}_{inepta}(\cdot \mid \cdot)$ can be transformed into $\mathbb{P}_{\text{subjective}}(\cdot \mid \cdot)$ simply by claiming that $\mathrm{P}_9(Red)$ and $\mathrm{P}_9(Green)$ represent his subjective degree of belief or betting propensities. Who can dispute it? The Bayesian is master of his own beliefs. However when elements from the two probability distributions that have such different meaning are mixed, the resultant number is just a number.

The Bayesian way around this problem is to rely of a higher power:

The Principal Principle:[13]

$$\mathbb{P}_{\text{subjective}}(a \mid \mathrm{P}_{objective}(a) = r) = r$$

This seems harmless enough as long as the parallel probability, $\mathbb{P}_{\text{subjective}}(\cdot \mid \cdot)$, is just used to guard against pairs of bets that are not sure to lose or to order his own thoughts. If the Bayesian asserts that his modified subjective probability represents a true statement about physical world, however conceived, extra assumptions are required. Another higher power:

The Lapicnirp Principle:

$$\mathrm{P}_{objective}(a \mid \mathbb{P}_{subjective}(a) = r) = r$$

[13]Howson and Urbach [**22**, p. 49] "The reasoning seems to rest on the rule of inference we referred to earlier, namely the ***principle of direct probability***, often called the ***principal principle***, which is used extensively in Bayesian statistics. It states that if the objective, physical probability of a random event (in the sense of its limiting relative frequency) were know to be r, and if no other relevant information were available, then the appropriate subjective degree of belief that the event will occur on any particular trial would also be r."

This is a very strange idea! Somehow the Bayesian must believe that his willingness to bet in a particular way causes truth to change about physical probability in the real world. Bayesians regularly act as though their having a strong opinion in favor of a proposition means that it is somehow close to being true, when in fact their actual probability statement merely reasserts that they are willing to bet or believe a certain way.

Subjective probability can never deliver ***true*** statements about the world we live in.

26.2. Convergence

The Relative *Odds* Stay Equal.

EXAMPLE 26.4.

Imagine $\widehat{\mathfrak{a}}$ and $\widehat{\mathfrak{b}}$, both Bayesian, *who believe that there are two types of urns*: *Green* ones with probabilities $\frac{1}{3}, \frac{1}{3}, \frac{1}{3}$ to produce *Black*, *Grey*, or *White* balls and *Red* ones with $\frac{1}{4}, \frac{1}{2}, \frac{1}{4}$ probability to produce *Black*, *Grey*, or *White* balls. The two differ on the probability of *Red* and *Green*. A number of balls are withdrawn from an urn of unknown color. This is e.

Bayesian theorists often argue that no matter what prior probabilities are chosen, as long as they are not one or zero, the steady collection of evidence will eventually produce posterior probabilities arbitrarily close to one answer.[14] This is true, but it is not clear that, that is enough. Here we will examine what happen to the odds in favor of an hypothesis when evidence accumulates.

THEOREM 26.1. **The collection of evidence makes no difference to the *relative* odds ratios of hypotheses in the Bayesian framework.**

Proof:

Consider the ratio between $\widehat{\mathfrak{a}}$'s and $\widehat{\mathfrak{b}}$'s views of the probability of *Red* vs. *Green* before any evidence has been collected.

$$\text{prior relative odds} = \frac{\left(\dfrac{\mathbb{P}_{\widehat{\mathfrak{a}}}(Red)}{\mathbb{P}_{\widehat{\mathfrak{a}}}(Green)}\right)}{\left(\dfrac{\mathbb{P}_{\widehat{\mathfrak{b}}}(Red)}{\mathbb{P}_{\widehat{\mathfrak{b}}}(Green)}\right)}$$

After the collection of evidence, e, by Bayes' theorem, assuming that none of the probabilities are zero, the similar ratio is:

$$\text{posterior relative odds} = \frac{\left(\dfrac{\mathbb{P}_{\mathfrak{a}}(Red|e)}{\mathbb{P}_{\mathfrak{a}}(Green|e)}\right)}{\left(\dfrac{\mathbb{P}_{\mathfrak{b}}(Red|e)}{\mathbb{P}_{\mathfrak{b}}(Green|e)}\right)}$$

[14] It is also true that no matter what the evidence, prior probabilities could have been chosen so skewed as to produce an arbitrarily incorrect answer.

$$\text{posterior relative odds} = \frac{\left(\dfrac{\left(\dfrac{\mathbb{P}_{\hat{\mathfrak{a}}}(Red)\mathbb{P}_{\hat{\mathfrak{a}}}(e|Red)}{\mathbb{P}_{\hat{\mathfrak{a}}}(e)}\right)}{\left(\dfrac{\mathbb{P}_{\hat{\mathfrak{a}}}(Green)\mathbb{P}_{\hat{\mathfrak{a}}}(e|Green)}{\mathbb{P}_{\hat{\mathfrak{a}}}(e)}\right)}\right)}{\left(\dfrac{\left(\dfrac{\mathbb{P}_{\hat{\mathfrak{b}}}(Red)\mathbb{P}_{\hat{\mathfrak{b}}}(e|Red)}{\mathbb{P}_{\hat{\mathfrak{b}}}(e)}\right)}{\left(\dfrac{\mathbb{P}_{\hat{\mathfrak{b}}}(Green)\mathbb{P}_{\hat{\mathfrak{b}}}(e|Green)}{\mathbb{P}_{\hat{\mathfrak{b}}}(e)}\right)}\right)}$$

Canceling $\mathbb{P}_{\hat{\mathfrak{a}}}(e)$ in the numerator and $\mathbb{P}_{\hat{\mathfrak{b}}}(e)$ in the denominator, we get:

$$= \frac{\left(\dfrac{\mathbb{P}_{\hat{\mathfrak{a}}}(Red)\mathbb{P}_{\hat{\mathfrak{a}}}(e|Red)}{\mathbb{P}_{\hat{\mathfrak{a}}}(Green)\mathbb{P}_{\hat{\mathfrak{a}}}(e|Green)}\right)}{\left(\dfrac{\mathbb{P}_{\hat{\mathfrak{b}}}(Red)\mathbb{P}_{\hat{\mathfrak{b}}}(e|Red)}{\mathbb{P}_{\hat{\mathfrak{b}}}(Green)\mathbb{P}_{\hat{\mathfrak{b}}}(e|Green)}\right)}$$

Then, because they are determined by the meaning of *Red* and *Green*

$$\mathbb{P}_{\hat{\mathfrak{a}}}(e \mid Red) = \mathbb{P}_{\hat{\mathfrak{b}}}(e \mid Red) \text{ and } \mathbb{P}_{\hat{\mathfrak{a}}}(e \mid Green) = \mathbb{P}_{\hat{\mathfrak{b}}}(e \mid Green),$$

$$\text{posterior relative odds} = \frac{\left(\dfrac{\mathbb{P}_{\hat{\mathfrak{a}}}(Red)}{\mathbb{P}_{\hat{\mathfrak{a}}}(Green)}\right)}{\left(\dfrac{\mathbb{P}_{\hat{\mathfrak{b}}}(Red)}{\mathbb{P}_{\hat{\mathfrak{b}}}(Green)}\right)}$$

Thus the ratio of posterior odds between the two distributions is identical to the ratio of prior odds. In other words, the accumulation of evidence leaves the divergence of opinion unchanged. *Q. E. D.*

26.3. God Creates the Universe

Suppose that God plays dice with the universe in a way more profound than Einstein could have imagined. Suppose that God collects all possible self-consistent laws of the universe, makes sealed copies of them in some proportion known only to Him and puts the copies in an urn. He then recruits Rev. Thomas Bayes to draw one envelope from the urn. God opens the envelope and installs the contents in the control box of the universe.

He turns to Bayes and says:

> In this laboratory are copies of all the possible self-consistent laws of the universe. Your are free to do any experiment you wish to gather evidence pertaining to which of the sets of laws are in place. After one hundred years, I require that you provide Me with posterior probabilities for each of the possible laws of the universe conditioned on the evidence you have collected.
>
> If what you provide Me is true, you go to Heaven; if it is not true, you go to Hell.

Bayes is doomed! His posterior probabilities can be true only if he can guess the true prior probabilities—if he can know the $\mathfrak{Mind\ of\ God}$. In order for his posterior probabilities to be *true*, $\mathrm{P}_{\mathfrak{God}}(\mathfrak{h})$ must equal $\mathbb{P}_{\widehat{Bayes}}(\mathfrak{h})$ for all possible self-consistent laws of the universe, $\mathfrak{h} \in \mathfrak{H}$.

$\mathfrak{God}$ is very tricky and has powers and abilities far beyond those of mortal men. There might be an infinite number of applicable laws and $\mathfrak{He}$ might have put one copy of each in the urn.[15] In that case, the natural choice is $\mathbb{P}_{\widehat{Bayes}}(\mathfrak{h}) = 0$ for all $\mathfrak{h}$. But if this is so, Bayes theorem cannot get initial traction.

$$\mathbb{P}(\mathfrak{h} \mid e) = \frac{\mathbb{P}(\mathfrak{h})\mathbb{P}(e|\mathfrak{h})}{(\mathbb{P}(\mathfrak{h})\mathbb{P}(e|\mathfrak{h})) + (\mathbb{P}(\neg\mathfrak{h})\mathbb{P}(e|\neg\mathfrak{h}))} = \frac{0}{1} = 0$$

On the other hand, $\mathfrak{He}$ might have played favorites and put one copy of one $\mathfrak{h}$ in the urn or, for that matter, any number of copies of a finite number of $\mathfrak{h}$. Then Rev. Bayes is faced with the problem of which $\mathfrak{h}_i$ to make $\mathbb{P}_{\widehat{Bayes}}(\mathfrak{h}_i) \neq 0$. If he assigns zero to an $\mathfrak{h}$ included in the urn or $\mathbb{P}_{\widehat{Bayes}}(\mathfrak{h}_i) > 0$ to an $\mathfrak{h}$ not in the the urn, he can never be even approximately correct.

If there are an infinite number of candidates, there is no hope; if there are a finite number of candidates, there are still an infinite number of possible of proportions, $\mathrm{P}_{\mathfrak{God}}(\mathfrak{h})$ that $\mathfrak{He}$ could have chosen. With collection of much evidence, if he knows what evidence to collect, Bayes can get closer to truth. However, if $\mathbb{P}_{\widehat{Bayes}}(\mathfrak{h}) \neq \mathrm{P}_{\mathfrak{God}}(\mathfrak{h})$, he can never get $\mathbb{P}_{\widehat{Bayes}}(\mathfrak{h} \mid e) = \mathrm{P}_{\mathfrak{God}}(\mathfrak{h} \mid e)$. Using Bayesian methods, truth cannot be found if it is not there at the beginning.

The best that Rev. Bayes can say is "I assert that I believe, to this extent, that $\mathbb{P}_{\widehat{Bayes}}(\mathfrak{h} \mid e)$ is true." $\mathfrak{God}$ doesn't care what Bayes believes. $\mathfrak{God}$ sends Bayes straight to $\mathfrak{Hell}$.

In situations less daunting than providing the laws of the universe, the same problems arise. The Bayesian can never assert the truth of a hypothesis; he can only assert what he believes. Who should care if $\mathfrak{God}$ doesn't?

The only escape is, if the prior probabilities are actual physical probabilities of an event and the Bayesian guesses correctly. Then and only then can a Bayesian assert truth rather than his belief.

The lesson here is that one ought to abandon the effort to find an unconditional probabilities for hypotheses and proceed with can be done. "Assuming one of these hypotheses is true, using the best decision rule available, based on this evidence, there is such and such a probability that *this* is the true one!"

Summary

There are many things that follow the probability axioms. If betting behavior is to be immune to Dutch book bet, it must follow the probability axioms. Subjective degree of belief is constructed to follow the probability axioms and must do so in order to avoid inconsistency. Silly things like, $\mathrm{P}_9(\cdot\,|\,\cdot)$, follow the probability axioms. Games of chance are hypothesized to follow the probability axioms. Errors of mathematical models in the soft sciences are hypothesized to follow the probability axioms.

Everyone, I think would that if an hypothesis predicts the evidence with a high likelihood, that should somehow be evidence that, that hypothesis is true; the question is, "How much?" Likelihoods do not follow the probability axioms. If one of $\mathfrak{h}_{1,\ldots,n} \in \mathfrak{H}$ is true,

$$\sum_{i=1}^{n} \mathrm{P}_{\mathfrak{h}_i}(e) \neq 1$$

The Bayesian supplies a rationale for making that sum equal one. By providing $\mathbb{P}(\mathfrak{h}_1), \ldots, \mathbb{P}(\mathfrak{h}_n)$, he has provided a *syntactic* means of normalizing the likelihoods.

The question is, "Are $\mathrm{P}_{\mathfrak{h}_i}(e)$, $\mathbb{P}_{\widehat{\mathfrak{b}}}(e \mid \mathfrak{h}_i)$, and $\mathbb{P}_{\widehat{\mathfrak{b}}}(\mathfrak{h}_i)$ made of the same stuff?" $\mathbb{P}_{\widehat{\mathfrak{b}}}(e \mid \mathfrak{h}_i)$ is asserted to be the same by the Principal Principle:

$$\mathrm{P}_{\mathfrak{h}_i}(e) = \mathbb{P}_{\widehat{\mathfrak{b}}}(e \mid \mathfrak{h}_i)$$

[15] Or any finite number of each, if the choices are infinite. Since the possibilities are not ordered, there is no way to arbitrarily assign prior probabilities as one could if they were points on a line.

Of course, this is perfectly proper. $\widehat{\mathfrak{b}}$ can choose his subjective probabilities any way he wants and can calculate a posterior probability for $\mathfrak{h}_i$ using Bayes' theorem:

$$\mathbb{P}_{\widehat{\mathfrak{b}}}(\mathfrak{h}_i \mid e) \propto \mathbb{P}_{\widehat{\mathfrak{b}}}(\mathfrak{h}_i) \times \mathbb{P}_{\widehat{\mathfrak{b}}}(e \mid \mathfrak{h}_i)$$

However, it remains a subjective, personal result. There is no interpersonal validity. If another Bayesian differs in assigning a prior probability to $\mathfrak{h}_i$, their posterior *odds* will never converge.

One could hypothesize that $\mathfrak{h}_i$ was chosen by a chance event and thus is of the same stuff as $\mathrm{P}_{\mathfrak{h}_i}(e)$. In that case, one has proposed a more complex probability space:

$$\mathrm{P}_{\mathfrak{h}'}(e \mid \mathfrak{h}_i) \propto \mathrm{P}_{\mathfrak{h}'}(\mathfrak{h}_i) \times \mathrm{P}_{\mathfrak{h}_i}(e)$$

It may be true or it may be false. It can be tested against alternative probability hypotheses. Were it true, based on e, one could calculate the probability that event, $\mathfrak{h}_i$, occurred. If a Bayesian guessed $\mathbb{P}_{\widehat{\mathfrak{b}}}(\mathfrak{h}_i) \neq \mathrm{P}_{\mathfrak{h}'}(\mathfrak{h}_i)$ used that prior for the purposes of making bets, he would, in some cases, accept losing bets and would accept progressively worse bets as evidence was collected.

If the Bayesian asserts that his prior was, *de facto*, the same stuff as $\mathrm{P}_{\mathfrak{h}_i}(e)$, he would be talking nonsense:

$$? \propto \mathbb{P}_{\widehat{\mathfrak{b}}}(\mathfrak{h}_i) \times \mathrm{P}_{\mathfrak{h}_i}(e)$$

Such a construction can be made syntactically correct, but it is as semantically empty as $\mathrm{P}_{inepta}(\cdot|\cdot)$ was. "What is 2 apples times 3 oranges?" Six what?

The Bayesian procedure delivers neither truth nor multi-party consensus.

Part VI

Conclusions

CHAPTER 27

What Has Been Shown

(1) Hume was right! Induction does not produce truth. The ashcan of science is full of inductively validated theories that were overturned by others that no one had thought of *yet*.

(2) Theories do not have *prior probabilities* unless they were in fact *chosen* by a probability process.

(3) If theories hypothesize the probability of evidence, it is possible to produce a statement, deductively true, in this form:

> If one of the theories is true, given the evidence and *this* decision rule, the probability of the decision rule producing the true theory is at least x.

(4) In most instances, there is a optimum decision rule that allows one to make simple statement of the probability that the hypothesis chosen by the decision rule is true, a statement that is deductively valid.

(5) One may wish for a procedure that produces an absolute probability that a hypothesis is true, but until one is found, one can *only* compare rivals. The hard sciences have done quite well with this procedure.

(6) Significance testing produces an unjustified sense of certainty. Odds of $9,999,999,999,999,999,999,999,999,999,999,999,999,999$ to 1 of correctly rejecting the the null hypothesis in a one-tailed t-test is the same as odds of 14.01315 to 1 that one would correctly choose a distribution like the sample over the null hypothesis using the optimum decision rule.

(7) All hypotheses validated by rejecting the null hypothesis using a significance test are suspect.

(8) When experiments in the soft sciences are replicated, they generally fail, even using the lax standards of significance testing. When hypotheses in the soft sciences are applied, they generally have lackluster results.

(9) In the soft sciences, simply by large sample size, approximately 100% of two-tailed tests and 50% of one-tailed tests will be found significant.

(10) It would be possible to reanalyze the data from past experiments and data-mining exercises, but, in general, only summary statistics are published and, thus, no reanalysis can be done.

(11) Twenty to fifty percent of what you and experts in the field believe to be true about diet, the causes of diseases, the effectiveness of medicines, how psychological factors affect behavior, the right way to treat social ills, the effects of economic policy, . . . are not true. [?]

Bibliography

1. Dante Alighieri, *Divina Commedia*, *Vulgari Eloquio* Press, Florence, c. 1300.
2. Fahiem Bacchus, *Representing and Reasoning with Probilististic Knowledge*, The MIT Press, Cambridge, Massachusetts-London, England, 1990.
3. Daryl J. Bem, *Feeling the Future: Experimental Evidence for Anomalous Retroactive Influences on Cognition and Affect*, Journal of Personality and Social Psychology **100** (2011), 407–425.
4. Daryl J. Bem, Jessica Utts, and Wesley O. Johnson, *Must Psychologists Change the Way They Analyse Their Data?*, Journal of Personality and Social Psychology **101** (2011), no. 4, 716–719.
5. Rudolf Carnap, *Logical Foundations of Probability*, second ed., The University of Chicago Press, Chicago, 1962.
6. P. H. Cootner (ed.), *The Random Character of Stock Market Prices*, MIT Press, Cambridge, Mass., 1964.
7. Bruno de Finetti, *Theory of Probability*, vol. 1, Wiley, London, 1974.
8. F. W. Dyson, A. S. Eddington, and C. Davidson, *A Determination of the Deflection of Light by the Sun's Graviational Field, from Obsevations Made at the Total Eclipse of May 29, 1919*, Memoirs of the Royal Astronomical Society **LLXII** (1923), 291–333.
9. J. Earman and C. Glymour, *Relativity and Eclipses: The Britsh Eclipse Expeditions of 1919 and Their Predecessors*, Historical Studies in the Physical Sciences **11** (1980), 49–85.
10. A. W. F. Edwards, *Likelihood*, expanded edition ed., Johns Hopkins University Press, Baltimore and London, 1992.
11. Albert Einstein, *Über den Einfluß der Schwercraft auf die Ausbreitung des Lichtes*, Annalen der Physik **35** (1911), 898–909.
12. ———, *Die Grundlagen der Allgemeinene Relativitätstheorie*, Annalen der Physik, Vierte Folge **49** (1916), no. 7, 769–822.
13. Errington et al, *An Open Investigation of the Reproducibility of Cancer Biology Research* , eLife **3** (2014), e04333.
14. L. Fisher and J. H. Lorie, *Rates of Return on Investments in Common Stock*, Journal of Business **37** (Jan., 1964), 1–17.
15. Ronald Aylmer Fisher, *Applications of "Student's" Distribution*, Metron **5** (1925), 90–104.
16. Gerd Gigerenzer, *Mindless Statistics*, The Journal of Socio-Economics **33(5)** (2004), 587–606.
17. A. J. Greenwald, *Consequences of Prejudice Against the Null Hypothesis*, Psychological Bulletin **82** (1975), 1–20.
18. Ian Hacking, *Logic of Statistical Inference*, Cambridge University Press, Cambridge, 1965.
19. Alan Hajek, *Scotching Dutch Books?*, Philosophical Perspectives **19** (2005), no. 1, 139–151.
20. Richard J. Herrnstein and Charles Murray, *The Bell Curve*, The Free Press, New York, 1994.
21. Colin Howson and Peter Urbach, *Scientific Reasoning: The Bayesian Approach*, Open Court, La Salle, Illinois, 1989.
22. ———, *Probability, Uncertainty and the Practice of Statistics*, Subjective Probability (George Wright and Peter Ayton, eds.), John Wiley & Sons, Chichester, 1994.
23. Roger G. Ibbotson and Gary P. Brinson, *Investment Markets: Gaining the Performance Advantage*, Mc Graw-Hill, New York, 1987.
24. John P. A. Ioannidis, *Why Most Published Research Findings Are False*, PLoS Med **2** (2005), no. 8, e124.
25. Daniel Kahneman and Amos Tversky, *Prospect Theory: An Analysis of Decision under Risk*, Econometrica **47** (1979), no. 2, 263–292.
26. Rex B. Kline, *Beyond Significance Testing: Statistics Reform in the Behavioral Sciences*, second ed., American Psychological Association, Washington, D.C., 2013.
27. A.N. Kolmogorov, *Grundbegriffe der Wahrscheinlichkeitsrechnung*, Springer, 1933.
28. ———, *Foundations of the Theory of Probability*, Chelsea, New York, 1950.
29. Arthur B. Laffer and R. David Ranson, *A Formal Model of the Economy*, The Journal of Business **44** (1971), no. 3, 247–270.
30. Jonah Lehrer, *The Truth Wears Off*, The New Yorker (December 13, 2010), 52–57.
31. D. V. Lindley, *Introduction Probability and Statistics from a Bayesian Viewpoint, Part 1: Probability*, Cambridge University Press, Cambridge, 1965.
32. Matthew C. Makel, Jonathan A. Plucker, and Boyd Hegarty, *Replications in Psychology Research: How Often Do They Really Occur?*, Science Perspectives on Psychological Science **7** (2012), no. 6, 537–542.
33. Paul E. Meehl, *Theory Testing in Psychology and Physics: A Methodological Paradox*, Philosophy of Science **34** (1967), 103–115.
34. ———, *Theoretical Risks and Tabular Asterisks: Sir Karl and Sir Ronald and The Slow Progress of Soft Psychology*, The Journal of Consulting and Clinical Psychology **46** (1978), 806–834.

35. Peter Milne, *Scotching the Dutch Book Argument*, Erkenntnis (1975-) **32** (1990), no. 1, 105–126.
36. Philip Morrison, Phylis Morrison, and the Office of Charles and Ray Eames, *Powers of Ten: About the Relative Size of Things in the Universe*, W. H. Freeman, San Francisco, 1982.
37. J. Neyman, *Outline of a Theory of Statistical Estimation Based on the Classical Theory of Probability*, Philosophical Transactions of the Royal Society of London. Series A, Mathematical and Physical Sciences **236** (1937), no. 767, pp. 333–380.
38. Brian A Nosek and Daniel Lakens, *Registered Reports: A Method to Increase the Credibility of Published Reports*, Social Psychology **45** (2014), no. 3, 137–141.
39. Karl Popper, *The Logic of Scientific Discovery*, Harper and Row, New York-Hagarstown-San Francisco-London, 1968.
40. S J Ritchie, R Wiseman, and C C French, *Failing the Future: Three Unsuccessful Attempts to Replicate Bem's 'Retroactive Facilitation of Recall' Effect*, PLoS ONE **7** (2010), e33423.
41. S. Roberts and H. Pashler, *How Persuasive Is a Good Fit? A Comment on Theory Testing*, Psychological Review **107** (2000), 358–367.
42. Jeffrey N. Rouder and Richard D. Morey, *A Bayes Factor Meta-Analysis of Bem's ESP Claim*, Psychonomic Bulletin & Review **18** (2011), 682–689.
43. Tina Hesman Saey, *Repeat Performance: Too Many Studies, When Replicated, Fail To Pass Muster*, Science News **187** (January 24, 2015), no. 2, 21–26.
44. Brian Skyrms, *Choice and Chance: An Introduction to Inductive Logic*, 2nd ed., Dickerson, Encino, California, 1975.
45. E. L. Smith, *Common Stocks as Long Term Investments*, Macmillan Co., New York, 1924.
46. Robert R. Sokal and F. James Rohlf, *Biometry*, 2nd ed., W. H. Freeman and Co., New York, 1981.
47. Amos Tversky, *Assessing Uncertainty*, Journal of the Royal Statistical Society. Series B (Methodological) **36** (1974), no. 2, 148–159.
48. Amos Tversky and Daniel Kahneman, *Judgment Under Uncertainty: Heuristics and Biases*, Science **185** (1974), no. 4157, 1124–1131.
49. ———, *Rational Choice in the Framing of Decisions*, The Journal of Business **59** (1986), no. 4, Part 2: The Behavioral Foundations of Economic Theory, S251–S278.
50. Eric Jan Wagenmakers, Ruud Wetzels, Denny Borsboom, and Han L. J. van der Maas, *Why Psychologists Must Change the Way They Analyze Their Data: The Case of PSI*, Journal of Personality and Social Psychology **100(3)** (Mar 2011), 426–432.
51. Niels G. Waller, *The Fallacy of the Null Hypothesis in Soft Psychology*, Applied & Preventive Psychology **11** (2004), 83–86.
52. Chamont Wang, *Sense and Nonsense of Statistical Inference: Controversy, Misuse, and Subtlety*, Marcel Dekker, New York-Basel-Hong Kong, 1993.
53. R Wetzels et al., *Statistical Evidence in Experimental Psychology: An Empirical Comparison Using 855 t Tests*, Perspectives on Psychological Science **6** (2011), no. 3, 291–298.

APPENDIX A

Popper's Probability Axioms

[?]

Set Notation

$a \in S$	means "a is an element of set S"
$S \cup T$	means "the set of all a such that $a \in S$ or $a \in T$"
$S \cap T$	means "the set of all a such that $a \in S$ and $a \in T$"

Why is *this* Axiomatization Important?

Most, if not all, discussions of probability use the axioms of A. Kolmogorov as a starting point [**27, 28**]. They are:
$P(a) \geq 0$
$P(a \cup b) = P(a) + P(b) - P(a \cap b)$
$P(\Omega) = 1$
where $P(\cdot)$ is a mapping of sets onto the real numbers greater than or equal to zero and less than or equal to one. $\{a, b, \dots\}$ are to be understood to be subsets of the universal set, Ω; $a \cup b$ is understood to be the union of a and b; and $a \cap b$ is the intersection of a and b. The *conditional probability of a given b* is defined in the following way:

$$P(a \mid b) = \frac{P(a \cap b)}{P(b)} \text{ if } P(b) \neq 0\text{; otherwise it does not exist.}$$

This is where things start to fall apart, if looked at closely. First, the usual formulation indicates that probabilities do not exist in situations where they are easily calculated. Consider the following example where it seems there is a probability distribution lurking in the background, but the definition of conditional probability blocks the way.

EXAMPLE A.1. Consider probability in the universe of positive integers. Let the probability be defined as:

$$\mathrm{P}(a) = \lim_{n\to\infty} \frac{\{\text{number of integers } \in a\} \leq n}{n}$$

Let r be the set of all prime numbers; let s be the set of every other prime starting with 1. $\mathrm{P}(r) = 0$, thus $\mathrm{P}(s \mid r)$ does not exist.

However jumping the tracks slightly, the natural solution is:

$$\mathrm{P}(s \mid r) = \lim_{n\to\infty} \frac{\{\text{every other prime}\} \leq n}{\{\text{every prime}\} \leq n} = \lim_{n\to\infty} \frac{\{\text{odd integers}\} \leq n}{\{\text{all integers}\} \leq n} = \frac{1}{2}$$

Here is something that looks like probability, but chokes on the Kolmogorov definition of conditional probability. What works in its original domain does not work everywhere.

There are yet other problems:

EXAMPLE A.2. Consider ordinary logic. Let $\mathrm{P}(true) = 1$ and $\mathrm{P}(false) = 0$.

Thus $\mathrm{P}(tautology) = 1$ and $\mathrm{P}(contradiction) = 0$.

However consider this: one wants $\mathrm{P}(a \mid b) = 1$, if b implies a. This is a problem because in ordinary first order logic, *contradiction* $\Rightarrow$ *anything* is **true**. This is because, since the contradiction can never be true, there can never be an occasion where the antecedent conditions exist to require the consequent to be true; thus the consequent can always be *true* or *false* and thus the implication is always true. To mimic ordinary logic one wants $\mathrm{P}(a \mid contradiction) = 1$ for any a, but in the Kolmogorov probability system, it not only is not equal to one, it does not exist.[1]

There is a way out of this muddle. About five years after [**27**] and continuing for another twenty years, Karl Popper, in a series of articles and appendices to *The Logic of Scientific Discovery* [**39**, appendices *II–*VII] produced an axiomatization of probability that solves all the problems inherent in the axioms in common use; no one followed his lead. His axioms have languished in *The Logic of Scientific Discovery*. It is a pity since they are correct where the usual system is wrong.

They have these advantages over the customary axioms:

- They are uninterpreted. Rather than defining probability as "subjective degree of belief" or "frequency in the long-run" or "propensity residing an experimental setup", or "measures on sets of real numbers." Popper's axioms provide only an algebra of arbitrary symbols.
- In Popper's system any seemingly well-formed formula, *is* a well-formed formula; if $\mathrm{P}(a \mid b)$ is meaningful, then $\mathrm{P}(b \mid a)$ is too. In the usual system it *might* be. If $\mathrm{P}(a) = 0$, it is meaningless.
- It automatically provide the structure of set theory or first order logic. A Boolean algebra is deducible from Popper's assignment of probability—the usual system is inconsistent with a Boolean algebra.
  - The usual axiomatizations cannot accommodate truth or falsehood. $\mathrm{P}(a) = 1$ won't do for truth; $\mathrm{P}(a) = 0$ won't do for falsehood.

    Neither will suffice because in the case of an infinite universe of discourse, there can be a finite or small infinity of contradictory cases.[2]

    The correct formulation is $\mathrm{P}(a \mid \neg a) = 0$ implies $\neg a$ is true and $\mathrm{P}(a \mid \neg a) = 1$ implies a is true.

---

[1]It must not be thought that these problems are merely formal. Researchers have be made grave mistakes by filling in these *lacunae* incorrectly. Consider the case in [**2**, p. 40] where the choice is made to assign $\mathrm{P}(a \mid b) = 0$ when $\mathrm{P}(b) = 0$. Bacchus inadvertently inserts a fatal flaw into his program because he has caused certain *tautologies* to have probability equal zero.

[2]Suppose an integer is drawn for a bag containing all positive integers. Most people would agree that $\mathrm{P}(42) = 0$ and $\mathrm{P}(\neg 42) = 1$. However, it is still possible that "42" could be drawn from the bag.

The usual axiomatization of probability forbids saying anything about $\mathrm{P}(\neg 42 \mid 42)$ if $\mathrm{P}(42) = 0$; it is undefined. In the following axiomatization, $\mathrm{P}(\neg 42 \mid 42) = 0$ if "42" is true and $\mathrm{P}(\neg 42 \mid 42) = 1$ if "42" is false. $\mathrm{P}(\neg a \mid a)$ always equals 0 or 1. Thus, completely within the probability calculus one can say a or $\neg a$ is the case. Nothing else is required.

- The usual axiomatizations cannot accommodate implication. $\mathrm{P}(a \mid b) = 1$ will not do for the same reason. The correct formulation is $\mathrm{P}(a \mid b\neg a) = 1$ This is the analog of $b \Rightarrow a$ in first-order logic.

These problems are all solved by the following axiom system.

Popper's Axioms

Logical Notation

$\exists a$	means "there exists an a such that..."
$\forall a$	means "for all a..."
$a \ \& \ b$	means "a and b"
$a \Rightarrow b$	means "if a, then b"
$a \not\Rightarrow b$	means "It is not the case that if a, then b"
$a \Leftrightarrow b$	means "a if and only if b"
$a \not\Leftrightarrow b$	means "It is not the case that a if and only if b"

Probability Notation

$\mathrm{P}(a \mid b)$	means "the probability of a given b"
$\mathrm{P}(a \mid b) = \mathrm{P}(c \mid d)$	means "$\mathrm{P}(a \mid b)$ is numerically equal to $\mathrm{P}(c \mid d)$"
$\mathrm{P}(a \mid b) \neq \mathrm{P}(c \mid d)$	means "$\mathrm{P}(a \mid b)$ is not numerically equal to $\mathrm{P}(c \mid d)$"
ab	is a primitive in the probability space. It can be interpreted as "a and b."[3]
$\neg a$	is a primitive in the probability space. It can be interpreted as "not a."[4]
$\neg ab$	can be interpreted as "b and not a."
$\neg(ab)$	can be interpreted "not the combination of a and b."
$a \vee b$	is a primitive in the probability space. It can be interpreted as "a or b."
$a \supseteq b$	is a primitive in the probability space. It can be interpreted as "b entails a."
$a \equiv b$	is a primitive in the probability space. It can be interpreted as "a is identical to b."

POSTULATE I. The number of elements in S is at most denumerably infinite.

Popper stated this postulate to show that it was never needed in the proofs that follow. In fact there *are* valid interpretations with more than countably many elements, but their explication requires extra terminology. Since the extension to *nondenumerable* interpretations is not needed in the proofs, it will be postponed to the end of this appendix.

POSTULATE II. If a and $b \in S$, then $\mathrm{P}(a \mid b)$ is a real number and the following hold:

AXIOM II.1. [Existence] $\exists c \exists d\, \mathrm{P}(a \mid b) \neq \mathrm{P}(c \mid d)$

AXIOM II.2. [Substitutivity] $\forall c\, \mathrm{P}(a \mid c) = \mathrm{P}(b \mid c) \Rightarrow \mathrm{P}(d \mid a) = \mathrm{P}(d \mid b)$

AXIOM II.3. [Reflexivity] $\mathrm{P}(a \mid a) = \mathrm{P}(b \mid b)$

POSTULATE III. If a and $b \in S$, then $ab \in S$; and if, moreover, $c \in S$ (and therefore bc) then the following hold:

AXIOM III.1. [Monotony] $\mathrm{P}(ab \mid c) \leq \mathrm{P}(a \mid c)$

AXIOM III.2. [Multiplication] $\mathrm{P}(ab \mid c) = \mathrm{P}(a \mid bc)\,\mathrm{P}(b \mid c)$

POSTULATE IV. If $a \in S$, then $\neg a \in S$; and if, moreover, $b \in S$, then the following holds:

AXIOM IV.1. [Complementation] $\mathrm{P}(a \mid a) \neq \mathrm{P}(b \mid a) \Rightarrow$
$\mathrm{P}(a \mid a) = \mathrm{P}(c \mid a) + \mathrm{P}(\neg c \mid a)$

POSTULATE V. [Absolute Probability] If a and $b \in S$, and if $\mathrm{P}(a \mid a) = \mathrm{P}(b \mid c)$ for every $c \in S$, then $\mathrm{P}(a) = \mathrm{P}(a \mid b)$

Upper & Lower Bounds

THEOREM A.1. $\mathrm{P}(a \mid a) = \mathrm{P}(b \mid b) = k$
Proof:
(Axiom II.3) $\mathrm{P}(a \mid a) = \mathrm{P}(b \mid b)$ [REFLEXIVITY] *Q. E. D.*

THEOREM A.2. $\mathrm{P}((aa)a \mid a) \leq \mathrm{P}(aa \mid a) \leq \mathrm{P}(a \mid a) = k$
Proof:
(Axiom III.1) $\mathrm{P}(ab \mid c) \leq \mathrm{P}(a \mid c)$ [MONOTONY]
(Thm. A.1) $\mathrm{P}(a \mid a) = \mathrm{P}(b \mid b) = k$ *Q. E. D.*

THEOREM A.3. $\mathrm{P}((aa)a \mid a) = \mathrm{P}(aa \mid aa)\,\mathrm{P}(a \mid a) = k^2$
Proof:
(Axiom III.2) $\mathrm{P}(ab \mid c) = \mathrm{P}(a \mid bc)\,\mathrm{P}(b \mid c)$ [MULTIPLICATION]
(Thm. A.1) $\mathrm{P}(a \mid a) = \mathrm{P}(b \mid b) = k$ *Q. E. D.*

THEOREM A.4. $k^2 \leq k$
Proof:
(Thm. A.2) $\mathrm{P}((aa)a \mid a) \leq \mathrm{P}(aa \mid a) \leq \mathrm{P}(a \mid a) = k$
(Thm. A.3) $\mathrm{P}((aa)a \mid a) = \mathrm{P}(aa \mid aa)\,\mathrm{P}(a \mid a) = k^2$ *Q. E. D.*

[3]These are natural interpretations that help in understanding what follows, but none of what follows depends on any particular interpretation. There are many other completely valid interpretations. One should think of this as moving symbols around on paper; interpretation follows.

[4]"¬" is commonly *defined* as a logical *not*, here it has a special meaning as defined in Postulate IV.

THEOREM A.5. $0 \leq k \leq 1$

Proof:

(Thm. A.4) $k^2 \leq k$

(Postulate II) *Q. E. D.*

THEOREM A.6. $k \neq \mathrm{P}(a \mid b) \Rightarrow k = k + \mathrm{P}(\neg b \mid b)$

Proof:

(Axiom IV.1) $\exists b \mathrm{P}(a \mid a) \neq \mathrm{P}(b \mid a) \Rightarrow \mathrm{P}(a \mid a) = \mathrm{P}(c \mid a) + \mathrm{P}(\neg c \mid a)$ [COMPLEMENTATION]

(Thm. A.1) $\mathrm{P}(a \mid a) = \mathrm{P}(b \mid b) = k$ *Q. E. D.*

THEOREM A.7. $k \neq \mathrm{P}(a \mid b) \Rightarrow \mathrm{P}(\neg b \mid b) = 0$

Proof:

(Thm. A.6) $k \neq \mathrm{P}(a \mid b) \Rightarrow k = k + \mathrm{P}(\neg b \mid b)$ *Q. E. D.*

THEOREM A.8. $\mathrm{P}(a\neg b \mid b) = \mathrm{P}(a \mid \neg bb)\,\mathrm{P}(\neg b \mid b)$

Proof:

(Axiom III.2) $\mathrm{P}(ab \mid c) = \mathrm{P}(a \mid bc)\,\mathrm{P}(b \mid c)$ [MULTIPLICATION] *Q. E. D.*

THEOREM A.9. $k \neq \mathrm{P}(a \mid b) \Rightarrow 0 = \mathrm{P}(a\neg b \mid b) \leq \mathrm{P}(a \mid b)$

Proof:

(Thm. A.7) $k \neq \mathrm{P}(a \mid b) \Rightarrow \mathrm{P}(\neg b \mid b) = 0$

(Thm. A.8) $\mathrm{P}(a\neg b \mid b) = \mathrm{P}(a \mid \neg bb)\,\mathrm{P}(\neg b \mid b)$

(Axiom III.1) $\mathrm{P}(ab \mid c) \leq \mathrm{P}(a \mid c)$ [MONOTONY] *Q. E. D.*

THEOREM A.10. $k \neq \mathrm{P}(a \mid b) \Rightarrow 0 \leq \mathrm{P}(a \mid b)$

Proof:

(Thm. A.9) $k \neq \mathrm{P}(a \mid b) \Rightarrow 0 = \mathrm{P}(a\neg b \mid b) \leq \mathrm{P}(a \mid b)$ *Q. E. D.*

THEOREM A.11. $0 > \mathrm{P}(a \mid b) \Rightarrow k = \mathrm{P}(a \mid b)$

Proof:

(Thm. A.10) $k \neq \mathrm{P}(a \mid b) \Rightarrow 0 \leq \mathrm{P}(a \mid b)$ *Q. E. D.*

THEOREM A.12. $k = \mathrm{P}(a \mid b) \Rightarrow 0 \leq \mathrm{P}(a \mid b)$

Proof:

(Thm. A.5) $0 \leq k \leq 1$ *Q. E. D.*

THEOREM A.13. $0 > \mathrm{P}(a \mid b) \Rightarrow 0 \leq \mathrm{P}(a \mid b)$

Proof:

(Thm. A.11) $0 > \mathrm{P}(a \mid b) \Rightarrow k = \mathrm{P}(a \mid b)$

(Thm. A.12) $k = \mathrm{P}(a \mid b) \Rightarrow 0 \leq \mathrm{P}(a \mid b)$ *Q. E. D.*

THEOREM A.14. $0 \leq \mathrm{P}(a \mid b)$

Proof:

(Thm. A.13) $0 > \mathrm{P}(a \mid b) \Rightarrow 0 \leq \mathrm{P}(a \mid b)$ *Q. E. D.*

THEOREM A.15. $0 \leq \mathrm{P}(\neg a \mid b)$

Proof:

(Thm. A.14) $0 \leq \mathrm{P}(a \mid b)$ *Q. E. D.*

THEOREM A.16. $k \neq \mathrm{P}(a \mid b) \Rightarrow k \geq \mathrm{P}(a \mid b)$

Proof:

(Axiom IV.1) $\exists b \mathrm{P}(a \mid a) \neq \mathrm{P}(b \mid a) \Rightarrow \mathrm{P}(a \mid a) = \mathrm{P}(c \mid a) + \mathrm{P}(\neg c \mid a)$ [COMPLEMENTATION]

(Thm. A.1) $\mathrm{P}(a \mid a) = \mathrm{P}(b \mid b) = k$

(Thm. A.15) $0 \leq \mathrm{P}(\neg a \mid b)$ *Q. E. D.*

THEOREM A.17. $\mathrm{P}(a \mid b) \leq k \leq 1$

Proof:

(Thm. A.16) $k \neq \mathrm{P}(a \mid b) \Rightarrow k \geq \mathrm{P}(a \mid b)$

(Thm. A.5) $0 \leq k \leq 1$ *Q. E. D.*

THEOREM A.18. $\mathbf{0 \leq P(a \mid b) \leq k \leq 1}$

Proof:

(Thm. A.14) $0 \leq \mathrm{P}(a \mid b)$

(Thm. A.17) $\mathrm{P}(a \mid b) \leq k \leq 1$ *Q. E. D.*

THEOREM A.19. $k = \mathrm{P}(aa \mid aa) \leq \mathrm{P}(a \mid aa) \leq k$

Proof:

(Thm. A.1) $\mathrm{P}(a \mid a) = \mathrm{P}(b \mid b) = k$

(Axiom III.1) $\mathrm{P}(ab \mid c) \leq \mathrm{P}(a \mid c)$ [MONOTONY]

(Thm. A.17) $\mathrm{P}(a \mid b) \leq k \leq 1$ *Q. E. D.*

THEOREM A.20. $k = \mathrm{P}(a(aa) \mid a(aa)) \leq \mathrm{P}(a \mid a(aa)) \leq k$

Proof:

(Thm. A.1) $\mathrm{P}(a \mid a) = \mathrm{P}(b \mid b) = k$

(Axiom III.1) $\mathrm{P}(ab \mid c) \leq \mathrm{P}(a \mid c)$ [MONOTONY]

(Thm. A.17) $\mathrm{P}(a \mid b) \leq k \leq 1$ *Q. E. D.*

THEOREM A.21. $k = \mathrm{P}(aa \mid aa) = \mathrm{P}(a \mid a(aa))\,\mathrm{P}(a \mid aa) = k^2$

Proof:

(Thm. A.1) $\mathrm{P}(a \mid a) = \mathrm{P}(b \mid b) = k$

(Axiom III.1) $\mathrm{P}(ab \mid c) \leq \mathrm{P}(a \mid c)$ [MONOTONY]

(Thm. A.19) $k = \mathrm{P}(aa \mid aa) \leq \mathrm{P}(a \mid aa) \leq k$

(Thm. A.20) $k = \mathrm{P}(a(aa) \mid a(aa)) \leq \mathrm{P}(a \mid a(aa)) \leq k$ *Q. E. D.*

THEOREM A.22. $k = k^2$

Proof:

(Thm. A.21) $k = \mathrm{P}(aa \mid aa) = \mathrm{P}(a \mid a(aa))\,\mathrm{P}(a \mid aa) = k^2$ *Q. E. D.*

THEOREM A.23. $\exists a \exists b \, \mathrm{P}(a \mid b) \neq 0 \Rightarrow k = 1$

Proof:

(Thm. A.18) $0 \leq \mathrm{P}(a \mid b) \leq k \leq 1$

(Thm. A.22) $k = k^2$ *Q. E. D.*

THEOREM A.24. $\exists a \exists b \, \mathrm{P}(a \mid b) \neq 0$

Proof:

(Axiom II.1) $\exists c \exists d \, \mathrm{P}(a \mid b) \neq \mathrm{P}(c \mid d)$ [EXISTENCE] *Q. E. D.*

THEOREM A.25. $\mathbf{P}(\boldsymbol{a} \mid \boldsymbol{a}) = \boldsymbol{k} = \mathbf{1}$

Proof:

(Thm. A.1) $\mathrm{P}(a \mid a) = \mathrm{P}(b \mid b) = k$

(Thm. A.23) $\exists a \exists b \, \mathrm{P}(a \mid b) \neq 0 \Rightarrow k = 1$

(Thm. A.24) $\exists a \exists b \, \mathrm{P}(a \mid b) \neq 0$ *Q. E. D.*

THEOREM A.26. $\exists b \exists a \, \mathrm{P}(b \mid a) \neq k$

Proof:

(Axiom II.1) $\exists c \exists d \, \mathrm{P}(a \mid b) \neq \mathrm{P}(c \mid d)$ [EXISTENCE]

(Thm. A.1) $\mathrm{P}(a \mid a) = \mathrm{P}(b \mid b) = k$ *Q. E. D.*

THEOREM A.27. $\exists \boldsymbol{a} \, \mathbf{P}(\neg \boldsymbol{a} \mid \boldsymbol{a}) = \mathbf{0}$

Proof:

(Thm. A.7) $k \neq \mathrm{P}(a \mid b) \Rightarrow \mathrm{P}(\neg b \mid b) = 0$

(Thm. A.26) $\exists b \exists a \, \mathrm{P}(b \mid a) \neq k$ *Q. E. D.*

Law of Idempotence

THEOREM A.28. $1 = \mathrm{P}(ab \mid ab) \leq \mathrm{P}(a \mid ab) = 1$

Proof:

(Thm. A.25) $\mathrm{P}(a \mid a) = k = 1$

(Axiom III.1) $\mathrm{P}(ab \mid c) \leq \mathrm{P}(a \mid c)$ [MONOTONY]

(Thm. A.17) $\mathrm{P}(a \mid b) \leq k \leq 1$ *Q. E. D.*

THEOREM A.29. $\mathrm{P}(aa \mid b) = \mathrm{P}(a \mid ab) \, \mathrm{P}(a \mid b)$

Proof:

(Axiom III.2) $\mathrm{P}(ab \mid c) = \mathrm{P}(a \mid bc) \, \mathrm{P}(b \mid c)$ [MULTIPLICATION] *Q. E. D.*

THEOREM A.30. **[Idempotence]** $\mathbf{P}(\boldsymbol{aa} \mid \boldsymbol{b}) = \mathbf{P}(\boldsymbol{a} \mid \boldsymbol{b})$

Proof:

(Thm. A.28) $1 = \mathrm{P}(ab \mid ab) \leq \mathrm{P}(a \mid ab) = 1$

(Thm. A.29) $\mathrm{P}(aa \mid b) = \mathrm{P}(a \mid ab) \, \mathrm{P}(a \mid b)$ *Q. E. D.*

Second Law of Monotony

THEOREM A.31. $\mathrm{P}(a \mid bc) < 1$

Proof:

(Thm. A.17) $\mathrm{P}(a \mid b) \leq k \leq 1$ *Q. E. D.*

THEOREM A.32. **[Second Law of Monotony]** $\mathbf{P}(ab \mid c) \leq \mathbf{P}(b \mid c)$
Proof:
(Axiom III.2) $\mathrm{P}(ab \mid c) = \mathrm{P}(a \mid bc)\mathrm{P}(b \mid c)$ [MULTIPLICATION]
(Thm. A.31) $\mathrm{P}(a \mid bc) \leq 1$
(Thm. A.14) $0 \leq \mathrm{P}(a \mid b)$ *Q. E. D.*

Commutation for First Argument

THEOREM A.33. $\mathrm{P}(a(bc) \mid a(bc)) = 1$
Proof:
(Thm. A.25) $\mathrm{P}(a \mid a) = k = 1$ *Q. E. D.*

THEOREM A.34. $\mathrm{P}(bc \mid a(bc)) = 1$
Proof:
(Thm. A.33) $\mathrm{P}(a(bc) \mid a(bc)) = 1$
(Thm. A.32) $\mathrm{P}(ab \mid c) \leq \mathrm{P}(b \mid c)$ [SECOND LAW OF MONOTONY]
(Thm. A.17) $\mathrm{P}(a \mid b) \leq k \leq 1$ *Q. E. D.*

THEOREM A.35. $\mathrm{P}(b \mid a(bc)) = 1$
Proof:
(Thm. A.34) $\mathrm{P}(bc \mid a(bc)) = 1$
(Axiom III.1) $\mathrm{P}(ab \mid c) \leq \mathrm{P}(a \mid c)$ [MONOTONY]
(Thm. A.17) $\mathrm{P}(a \mid b) \leq k \leq 1$ *Q. E. D.*

THEOREM A.36. $\mathrm{P}(ba \mid bc) = \mathrm{P}(a \mid bc)$
Proof:
(Thm. A.35) $\mathrm{P}(b \mid a(bc)) = 1$
(Axiom III.2) $\mathrm{P}(ab \mid c) = \mathrm{P}(a \mid bc)\mathrm{P}(b \mid c)$ [MULTIPLICATION] *Q. E. D.*

THEOREM A.37. $\mathrm{P}((ba)b \mid c) = \mathrm{P}(ab \mid c)$
Proof:
(Thm. A.36) $\mathrm{P}(ba \mid bc) = \mathrm{P}(a \mid bc)$
(Axiom III.2) $\mathrm{P}(ab \mid c) = \mathrm{P}(a \mid bc)\mathrm{P}(b \mid c)$ [MULTIPLICATION] *Q. E. D.*

THEOREM A.38. $\mathrm{P}(ba \mid c) \geq \mathrm{P}(ab \mid c)$
Proof:
(Thm. A.37) $\mathrm{P}((ba)b \mid c) = \mathrm{P}(ab \mid c)$
(Axiom III.1) $\mathrm{P}(ab \mid c) \leq \mathrm{P}(a \mid c)$ [MONOTONY] *Q. E. D.*

THEOREM A.39. $\mathrm{P}(ab \mid c) \geq \mathrm{P}(ba \mid c)$
Proof:
(Thm. A.38) $\mathrm{P}(ba \mid c) \geq \mathrm{P}(ab \mid c)$ *Q. E. D.*

THEOREM A.40. **[Commutation of First Argument]**
$\mathbf{P}(ab \mid c) = \mathbf{P}(ba \mid c)$

Proof:

(Thm. A.38) $\mathrm{P}(ba \mid c) \geq \mathrm{P}(ab \mid c)$

(Thm. A.39) $\mathrm{P}(ab \mid c) \geq \mathrm{P}(ba \mid c)$ *Q. E. D.*

Association for First Argument

THEOREM A.41. $\mathrm{P}(ab \mid d((ab)c)) = 1$

Proof:

(Thm. A.35) $\mathrm{P}(b \mid a(bc)) = 1$ *Q. E. D.*

THEOREM A.42. $\mathrm{P}(a \mid d((ab)c)) = 1 = \mathrm{P}(b \mid d((ab)c))$

Proof:

(Thm. A.41) $\mathrm{P}(ab \mid d((ab)c)) = 1$

(Axiom III.1) $\mathrm{P}(ab \mid c) \leq \mathrm{P}(a \mid c)$ [MONOTONY]

(Thm. A.17) $\mathrm{P}(a \mid b) \leq k \leq 1$

(Thm. A.32) $\mathrm{P}(ab \mid c) \leq \mathrm{P}(b \mid c)$ [SECOND LAW OF MONOTONY] *Q. E. D.*

THEOREM A.43. $\mathrm{P}(a \mid (bc)((ab)c)) = 1$

Proof:

(Thm. A.42) $\mathrm{P}(a \mid d((ab)c)) = 1 = \mathrm{P}(b \mid d((ab)c))$ *Q. E. D.*

THEOREM A.44. $\mathrm{P}(a(bc) \mid (ab)c) = \mathrm{P}(bc \mid (ab)c)$

Proof:

(Thm. A.43) $\mathrm{P}(a \mid (bc)((ab)c)) = 1$

(Axiom III.2) $\mathrm{P}(ab \mid c) = \mathrm{P}(a \mid bc)\,\mathrm{P}(b \mid c)$ [MULTIPLICATION] *Q. E. D.*

THEOREM A.45. $\mathrm{P}(bc \mid (ab)c) = \mathrm{P}(b \mid c((ab)c))\,\mathrm{P}(c \mid (ab)c)$

Proof:

(Axiom III.2) $\mathrm{P}(ab \mid c) = \mathrm{P}(a \mid bc)\,\mathrm{P}(b \mid c)$ [MULTIPLICATION] *Q. E. D.*

THEOREM A.46. $\mathrm{P}(b \mid c((ab)c)) = 1$

Proof:

(Thm. A.42) $\mathrm{P}(a \mid d((ab)c)) = 1 = \mathrm{P}(b \mid d((ab)c))$ *Q. E. D.*

THEOREM A.47. $\mathrm{P}(c \mid (ab)c) = 1$

Proof:

(Thm. A.25) $\mathrm{P}(a \mid a) = k = 1$

(Thm. A.32) $\mathrm{P}(ab \mid c) \leq \mathrm{P}(b \mid c)$ [SECOND LAW OF MONOTONY]

(Thm. A.17) $\mathrm{P}(a \mid b) \leq k \leq 1$ *Q. E. D.*

THEOREM A.48. $\mathrm{P}(a(bc) \mid (ab)c) = 1$

Proof:

(Thm. A.44) $\mathrm{P}(a(bc) \mid (ab)c) = \mathrm{P}(bc \mid (ab)c)$

(Thm. A.45) $\mathrm{P}(bc \mid (ab)c) = \mathrm{P}(b \mid c((ab)c))\,\mathrm{P}(c \mid (ab)c)$

(Thm. A.46) $\mathrm{P}(b \mid c((ab)c)) = 1$

(Thm. A.47) $\mathrm{P}(c \mid (ab)c) = 1$ *Q. E. D.*

THEOREM A.49. $\mathrm{P}(a(b(cd)) \mid d) = \mathrm{P}(cd \mid b(ad))\,\mathrm{P}(b \mid ad)\,\mathrm{P}(a \mid d)$

Proof:

(Thm. A.40) $\mathrm{P}(ab \mid c) = \mathrm{P}(ba \mid c)$
[COMMUTATION OF FIRST ARGUMENT]

(Axiom III.2) $\mathrm{P}(ab \mid c) = \mathrm{P}(a \mid bc)\,\mathrm{P}(b \mid c)$ [MULTIPLICATION] *Q. E. D.*

THEOREM A.50. $\mathrm{P}(a(bc) \mid d) = \mathrm{P}(c \mid b(ad))\,\mathrm{P}(b \mid ad)\,\mathrm{P}(a \mid d)$

Proof:

(Thm. A.40) $\mathrm{P}(ab \mid c) = \mathrm{P}(ba \mid c)$
[COMMUTATION OF FIRST ARGUMENT]

(Axiom III.2) $\mathrm{P}(ab \mid c) = \mathrm{P}(a \mid bc)\,\mathrm{P}(b \mid c)$ [MULTIPLICATION] *Q. E. D.*

THEOREM A.51. $\mathrm{P}(a(bc) \mid d) \geq \mathrm{P}(a(b(cd)) \mid d)$

Proof:

(Thm. A.49) $\mathrm{P}(a(b(cd)) \mid d) = \mathrm{P}(cd \mid b(ad))\,\mathrm{P}(b \mid ad)\,\mathrm{P}(a \mid d)$

(Thm. A.50) $\mathrm{P}(a(bc) \mid d) = \mathrm{P}(c \mid b(ad))\,\mathrm{P}(b \mid ad)\,\mathrm{P}(a \mid d)$

(Axiom III.1) $\mathrm{P}(ab \mid c) \leq \mathrm{P}(a \mid c)$ [MONOTONY] *Q. E. D.*

THEOREM A.52. $\mathrm{P}(a(b(cd)) \mid (ab)(cd)) = 1$

Proof:

(Thm. A.48) $\mathrm{P}(a(bc) \mid (ab)c) = 1$ *Q. E. D.*

THEOREM A.53. $\mathrm{P}((a(b(cd))(ab) \mid cd) = \mathrm{P}(ab \mid cd)$

Proof:

(Thm. A.52) $\mathrm{P}(a(b(cd)) \mid (ab)(cd)) = 1$

(Axiom III.2) $\mathrm{P}(ab \mid c) = \mathrm{P}(a \mid bc)\,\mathrm{P}(b \mid c)$ [MULTIPLICATION] *Q. E. D.*

THEOREM A.54. $\mathrm{P}(a(b(cd)) \mid cd) \geq \mathrm{P}(ab \mid cd)$

Proof:

(Thm. A.53) $\mathrm{P}((a(b(cd))(ab) \mid cd) = \mathrm{P}(ab \mid cd)$

(Axiom III.1) $\mathrm{P}(ab \mid c) \leq \mathrm{P}(a \mid c)$ [MONOTONY] *Q. E. D.*

THEOREM A.55. $\mathrm{P}((a(b(cd)))c \mid d) \geq \mathrm{P}((ab)c \mid d)$

Proof:

(Thm. A.54) $\mathrm{P}(a(b(cd)) \mid cd) \geq \mathrm{P}(ab \mid cd)$

(Axiom III.2) $\mathrm{P}(ab \mid c) = \mathrm{P}(a \mid bc)\,\mathrm{P}(b \mid c)$ [MULTIPLICATION] *Q. E. D.*

THEOREM A.56. $\mathrm{P}(a(b(cd)) \mid d) \geq \mathrm{P}((ab)c \mid d)$

Proof:

(Thm. A.55) $\mathrm{P}((a(b(cd)))c \mid d) \geq \mathrm{P}((ab)c \mid d)$

(Axiom III.1) $\mathrm{P}(ab \mid c) \leq \mathrm{P}(a \mid c)$ [MONOTONY] *Q. E. D.*

THEOREM A.57. $\mathrm{P}(a(bc) \mid d) \geq \mathrm{P}((ab)c \mid d)$

Proof:

(Thm. A.51) $\mathrm{P}(a(bc) \mid d) \geq \mathrm{P}(a(b(cd)) \mid d)$

(Thm. A.56) $\mathrm{P}(a(b(cd)) \mid d) \geq \mathrm{P}((ab)c \mid d)$ *Q. E. D.*

THEOREM A.58. $\mathrm{P}((bc)a \mid d) \geq \mathrm{P}((ab)c \mid d)$

Proof:

(Thm. A.57) $\mathrm{P}(a(bc) \mid d) \geq \mathrm{P}((ab)c \mid d)$

(Thm. A.40) $\mathrm{P}(ab \mid c) = \mathrm{P}(ba \mid c)$
[COMMUTATION OF FIRST ARGUMENT] *Q. E. D.*

THEOREM A.59. $\mathrm{P}((ab)c \mid d) \geq \mathrm{P}(b(ca) \mid d)$

Proof:

(Thm. A.58) $\mathrm{P}((bc)a \mid d) \geq \mathrm{P}((ab)c \mid d)$

(Thm. A.40) $\mathrm{P}(ab \mid c) = \mathrm{P}(ba \mid c)$
[COMMUTATION OF FIRST ARGUMENT] *Q. E. D.*

THEOREM A.60. $\mathrm{P}((bc)a \mid d) \geq \mathrm{P}(b(ca) \mid d)$

Proof:

(Thm. A.58) $\mathrm{P}((bc)a \mid d) \geq \mathrm{P}((ab)c \mid d)$

(Thm. A.59) $\mathrm{P}((ab)c \mid d) \geq \mathrm{P}(b(ca) \mid d)$ *Q. E. D.*

THEOREM A.61. $\mathrm{P}((ab)c \mid d) \geq \mathrm{P}(a(bc) \mid d)$

Proof:

(Thm. A.60) $\mathrm{P}((bc)a \mid d) \geq \mathrm{P}(b(ca) \mid d)$ *Q. E. D.*

THEOREM A.62. **[Association for First Argument]**
$\mathbf{P}((\boldsymbol{ab})\boldsymbol{c} \mid \boldsymbol{d}) = \mathbf{P}(\boldsymbol{a}(\boldsymbol{bc}) \mid \boldsymbol{d})$

Proof:

(Axiom II.2) $\forall c \mathrm{P}(a \mid c) = \mathrm{P}(b \mid c) \Rightarrow \mathrm{P}(d \mid a) = \mathrm{P}(d \mid b)$
[SUBSTITUTIVITY] *Q. E. D.*

Complementation

THEOREM A.63. $\mathrm{P}(\neg b \mid b) \neq 0 \Leftrightarrow \forall c \mathrm{P}(c \mid b) = 1$

Proof:

(Thm. A.7) $k \neq \mathrm{P}(a \mid b) \Rightarrow \mathrm{P}(\neg b \mid b) = 0$

(Thm. A.25) $\mathrm{P}(a \mid a) = k = 1$ *Q. E. D.*

THEOREM A.64. **[Unconditional form of complementation]** $\mathbf{P(a \mid b) + P(\neg a \mid b) = 1 + P(\neg b \mid b)}$

Proof:

(Axiom IV.1) $\exists b \, \mathrm{P}(a \mid a) \neq \mathrm{P}(b \mid a) \Rightarrow \mathrm{P}(a \mid a) = \mathrm{P}(c \mid a) + \mathrm{P}(\neg c \mid a)$ [COMPLEMENTATION]

(Thm. A.25) $\mathrm{P}(a \mid a) = k = 1$

(Thm. A.63) $\mathrm{P}(\neg b \mid b) \neq 0 \Leftrightarrow \forall c \, \mathrm{P}(c \mid b) = 1$ *Q. E. D.*

THEOREM A.65. $\mathrm{P}(a \mid b) + \mathrm{P}(\neg a \mid b) = \mathrm{P}(c \mid b) + \mathrm{P}(\neg c \mid b)$

Proof:

(Thm. A.64) $\mathrm{P}(a \mid b) + \mathrm{P}(\neg a \mid b) = 1 + \mathrm{P}(\neg b \mid b)$ *Q. E. D.*

THEOREM A.66. $\mathrm{P}(a \mid bd) + \mathrm{P}(\neg a \mid bd) = \mathrm{P}(c \mid bd) + \mathrm{P}(\neg c \mid bd)$

Proof:

(Thm. A.65) $\mathrm{P}(a \mid b) + \mathrm{P}(\neg a \mid b) = \mathrm{P}(c \mid b) + \mathrm{P}(\neg c \mid b)$ *Q. E. D.*

THEOREM A.67. $\mathrm{P}(ab \mid d) + \mathrm{P}(\neg ab \mid d) = \mathrm{P}(cb \mid d) + \mathrm{P}(\neg cb \mid d)$

Proof:

(Axiom III.2) $\mathrm{P}(ab \mid c) = \mathrm{P}(a \mid bc) \mathrm{P}(b \mid c)$ [MULTIPLICATION]

(Thm. A.66) $\mathrm{P}(a \mid bd) + \mathrm{P}(\neg a \mid bd) = \mathrm{P}(c \mid bd) + \mathrm{P}(\neg c \mid bd)$ *Q. E. D.*

THEOREM A.68. $\mathrm{P}(ab \mid c) + \mathrm{P}(\neg ab \mid c) = \mathrm{P}(cb \mid c) + \mathrm{P}(\neg cb \mid c)$

Proof:

(Thm. A.67) $\mathrm{P}(ab \mid d) + \mathrm{P}(\neg ab \mid d) = \mathrm{P}(cb \mid d) + \mathrm{P}(\neg cb \mid d)$ *Q. E. D.*

THEOREM A.69. $\mathrm{P}(\neg cb \mid c) = \mathrm{P}(\neg c \mid c)$

Proof:

(Thm. A.7) $k \neq \mathrm{P}(a \mid b) \Rightarrow \mathrm{P}(\neg b \mid b) = 0$

(Axiom III.1) $\mathrm{P}(ab \mid c) \leq \mathrm{P}(a \mid c)$ [MONOTONY]

(Thm. A.25) $\mathrm{P}(a \mid a) = k = 1$

(Thm. A.63) $\mathrm{P}(\neg b \mid b) \neq 0 \Leftrightarrow \forall c \, \mathrm{P}(c \mid b) = 1$ *Q. E. D.*

THEOREM A.70. $\mathrm{P}(ab \mid ab) = 1 = \mathrm{P}(b \mid ab)$

Proof:

(Thm. A.25) $\mathrm{P}(a \mid a) = k = 1$

(Thm. A.32) $\mathrm{P}(ab \mid c) \leq \mathrm{P}(b \mid c)$ [SECOND LAW OF MONOTONY]

(Thm. A.17) $\mathrm{P}(a \mid b) \leq k \leq 1$ *Q. E. D.*

THEOREM A.71. $\mathrm{P}(ba \mid b) = \mathrm{P}(b \mid ab) \mathrm{P}(a \mid b) = \mathrm{P}(a \mid b)$

Proof:

(Axiom III.2) $\mathrm{P}(ab \mid c) = \mathrm{P}(a \mid bc) \mathrm{P}(b \mid c)$ [MULTIPLICATION]

(Thm. A.70) $\mathrm{P}(ab \mid ab) = 1 = \mathrm{P}(b \mid ab)$ *Q. E. D.*

THEOREM A.72. $P(ab \mid b) = P(a \mid bb) = P(a \mid b)$

Proof:

(Axiom III.2)	$P(ab \mid c) = P(a \mid bc) P(b \mid c)$ [MULTIPLICATION]
(Thm. A.25)	$P(a \mid a) = k = 1$
(Thm. A.71)	$P(ba \mid b) = P(b \mid ab) P(a \mid b) = P(a \mid b)$
(Thm. A.40)	$P(ab \mid c) = P(ba \mid c)$ [COMMUTATION OF FIRST ARGUMENT] *Q. E. D.*

The following six theorems are not true in the usual systems. They are necessary if probability is to be consistent with ordinary first order logic. The first five show that $P(a \mid b) = 1$ if b is a contradiction. The sixth elaborates the result of theorem A.64.

THEOREM A.73. $\mathbf{P(a \mid a\neg a) = 1}$

Proof:

(Thm. A.28) $1 = P(ab \mid ab) \leq P(a \mid ab) = 1$ *Q. E. D.*

THEOREM A.74. $\mathbf{P(\neg a \mid a\neg a) = 1}$

Proof:

(Thm. A.28) $1 = P(ab \mid ab) \leq P(a \mid ab) = 1$ *Q. E. D.*

THEOREM A.75. $\mathbf{P(a \mid \neg bb) = 1}$

Proof:

(Thm. A.73)	$P(a \mid a\neg a) = 1$
(Thm. A.74)	$P(\neg a \mid a\neg a) = 1$
(Axiom IV.1)	$\exists b P(a \mid a) \neq P(b \mid a) \Rightarrow P(a \mid a) = P(c \mid a) + P(\neg c \mid a)$ [COMPLEMENTATION] *Q. E. D.*

THEOREM A.76. $\mathbf{\exists b \forall a P(a \mid b) = 1}$

Proof:

(Thm. A.75) $P(a \mid \neg bb) = 1$ *Q. E. D.*

THEOREM A.77. $\mathbf{\exists a P(\neg a \mid a) = 1}$

Proof:

(Thm. A.76) $\exists b \forall a P(a \mid b) = 1$ *Q. E. D.*

THEOREM A.78. $\mathbf{P(ab \mid c) + P(\neg ab \mid c) = P(b \mid c) + P(\neg c \mid c)}$

Proof:

(Thm. A.68)	$P(ab \mid c) + P(\neg ab \mid c) = P(cb \mid c) + P(\neg cb \mid c)$
(Thm. A.69)	$P(\neg cb \mid c) = P(\neg c \mid c)$
(Thm. A.71)	$P(ba \mid b) = P(b \mid ab) P(a \mid b) = P(a \mid b)$ *Q. E. D.*

This theorem and theorem A.64 show that for consistency the conditional probability of a statement and its negation must total 2 if conditioning statement is a contradiction. This need cause no distress because since the conditioning statement can never be true, the conditional probability never reaches the real world, so to speak. It mimics the situation in logic where $a \Rightarrow b$ is always true, no matter what, if a is a contradiction. If $P(c) \neq 0$, then $P(\neg c \mid c) = 0$ and the formula of the ordinary system obtains.

Absolute Probability

THEOREM A.79. $P(aa \mid b) + P(\neg aa \mid b) = P(a \mid b) + P(\neg b \mid b)$

Proof:

(Thm. A.78) $P(ab \mid c) + P(\neg ab \mid c) = P(b \mid c) + P(\neg c \mid c)$ *Q. E. D.*

THEOREM A.80. $P(\neg aa \mid b) = P(a\neg a \mid b) = P(\neg b \mid b)$

Proof:

(Thm. A.40) $P(ab \mid c) = P(ba \mid c)$
[COMMUTATION OF FIRST ARGUMENT]

(Thm. A.79) $P(aa \mid b) + P(\neg aa \mid b) = P(a \mid b) + P(\neg b \mid b)$

(Thm. A.30) $P(aa \mid b) = P(a \mid b)$ [IDEMPOTENCE] *Q. E. D.*

THEOREM A.81. $P(\neg aa \mid b) + P(\neg(\neg aa) \mid b) = P(a\neg a \mid b) + P(\neg(a\neg a) \mid b) = 1 + P(\neg b \mid b)$

Proof:

(Thm. A.64) $P(a \mid b) + P(\neg a \mid b) = 1 + P(\neg b \mid b)$ *Q. E. D.*

THEOREM A.82. **$P(\neg(\neg aa) \mid b) = 1 = P(\neg(a\neg a) \mid b)$**
["$\neg(a\neg a)$" satisfies Postulate V]

Proof:

(Thm. A.80) $P(\neg aa \mid b) = P(a\neg a \mid b) = P(\neg b \mid b)$

(Thm. A.81) $P(\neg aa \mid b) + P(\neg(\neg aa) \mid b) = P(a\neg a \mid b) + P(\neg(a\neg a) \mid b) = 1 + P(\neg b \mid b)$ *Q. E. D.*

THEOREM A.83. **$P(a) = P(a \mid \neg(a\neg a)) = P(a \mid \neg(\neg aa)) = P(a \mid \neg(b\neg b)) = P(a \mid \neg(\neg bb))$**

Proof:

(Thm. A.25) $P(a \mid a) = k = 1$

(Thm. A.82) $P(\neg(\neg aa) \mid b) = 1 = P(\neg(a\neg a) \mid b)$

(Postulate V) If a and $b \in S$, and if $P(a \mid a) = P(b \mid c)$ for every $c \in S$, then $P(a) = P(a \mid b)$ *Q. E. D.*

S Is Necessarily a Boolean Algebra

There is a clash between the ordinary axiomatization and first order logic. To show this, in this section, we show that the assignment of probabilities automatically produces a structure that is a Boolean algebra. In the case of statements, this produces the usual structure of *and, or, not, implication,* and *logical equivalence* with *true* implying probability one and *false* implying probability zero.

THEOREM A.84. $P(a\neg b \mid c) = P(a \mid c) - P(ab \mid c) + P(\neg c \mid c)$

Proof:

(Thm. A.78) $P(ab \mid c) + P(\neg ab \mid c) = P(b \mid c) + P(\neg c \mid c)$

(Thm. A.40) $P(ab \mid c) = P(ba \mid c)$
[COMMUTATION OF FIRST ARGUMENT] *Q. E. D.*

THEOREM A.85. $P(\neg a\neg b \mid c) = P(\neg a \mid c) - P(\neg ab \mid c) + P(\neg c \mid c)$

Proof:

(Thm. A.84) $P(a\neg b \mid c) = P(a \mid c) - P(ab \mid c) + P(\neg c \mid c)$ *Q. E. D.*

THEOREM A.86. $\mathrm{P}(\neg a \neg b \mid c) = 1 - \mathrm{P}(a \mid c) - \mathrm{P}(b \mid c) + \mathrm{P}(ab \mid c) + \mathrm{P}(\neg c \mid c)$

Proof:

(Thm. A.85) $\mathrm{P}(\neg a \neg b \mid c) = \mathrm{P}(\neg a \mid c) - \mathrm{P}(\neg ab \mid c) + \mathrm{P}(\neg c \mid c)$

(Thm. A.84) $\mathrm{P}(a \neg b \mid c) = \mathrm{P}(a \mid c) - \mathrm{P}(ab \mid c) + \mathrm{P}(\neg c \mid c)$

(Thm. A.64) $\mathrm{P}(a \mid b) + \mathrm{P}(\neg a \mid b) = 1 + \mathrm{P}(\neg b \mid b)$

(Thm. A.40) $\mathrm{P}(ab \mid c) = \mathrm{P}(ba \mid c)$
[COMMUTATION OF FIRST ARGUMENT] *Q. E. D.*

DEFINITION A.1. **[Equivalence]** $\boldsymbol{a \equiv b \Leftrightarrow \forall x\, \mathrm{P}(a \mid x) = \mathrm{P}(b \mid x)}$

THEOREM A.87. $a \equiv a$

THEOREM A.88. $a \equiv b \Rightarrow b \equiv a$

THEOREM A.89. $(a \equiv b)\,\&\,(b \equiv c) \Rightarrow a \equiv c$

DEFINITION A.2. **[Boolean Sum]** $\boldsymbol{a \equiv b \vee c \Leftrightarrow a \equiv \neg(\neg b \neg c)}$

THEOREM A.90. **["$\boldsymbol{\mathrm{P}(a \vee b \mid c) =}$"]**
$\boldsymbol{\mathrm{P}(\neg(\neg a \neg b) \mid c) = \mathrm{P}(a \mid c) + \mathrm{P}(b \mid c) - \mathrm{P}(ab \mid c)}$

Proof:

(Thm. A.86) $\mathrm{P}(\neg a \neg b \mid c) = 1 - \mathrm{P}(a \mid c) - \mathrm{P}(b \mid c) + \mathrm{P}(ab \mid c) + \mathrm{P}(\neg c \mid c)$

(Thm. A.64) $\mathrm{P}(a \mid b) + \mathrm{P}(\neg a \mid b) = 1 + \mathrm{P}(\neg b \mid b)$ *Q. E. D.*

THEOREM A.91. $\mathrm{P}(\neg(\neg b \neg c) \mid ad) = \mathrm{P}(b \mid ad) + \mathrm{P}(c \mid ad) - \mathrm{P}(bc \mid ad)$

Proof:

(Thm. A.90) $\mathrm{P}(\neg(\neg a \neg b) \mid c) = \mathrm{P}(a \mid c) + \mathrm{P}(b \mid c) - \mathrm{P}(ab \mid c)$ *Q. E. D.*

THEOREM A.92. $\boldsymbol{\mathrm{P}(a \neg(\neg b \neg c) \mid d) = \mathrm{P}(ab \mid d) + \mathrm{P}(ac \mid d) - \mathrm{P}(a(bc) \mid d)}$

Proof:

(Thm. A.91) $\mathrm{P}(\neg(\neg b \neg c) \mid ad) = \mathrm{P}(b \mid ad) + \mathrm{P}(c \mid ad) - \mathrm{P}(bc \mid ad)$

(Axiom III.2) $\mathrm{P}(ab \mid c) = \mathrm{P}(a \mid bc)\,\mathrm{P}(b \mid c)$ [MULTIPLICATION]

(Thm. A.40) $\mathrm{P}(ab \mid c) = \mathrm{P}(ba \mid c)$
[COMMUTATION OF FIRST ARGUMENT] *Q. E. D.*

THEOREM A.93. $\mathrm{P}(a(bc) \mid d) = \mathrm{P}(a \mid (bc)d)\,\mathrm{P}(bc \mid d) = \mathrm{P}((aa)(bc) \mid d)$

Proof:

(Axiom III.2) $\mathrm{P}(ab \mid c) = \mathrm{P}(a \mid bc)\,\mathrm{P}(b \mid c)$ [MULTIPLICATION]

(Thm. A.30) $\mathrm{P}(aa \mid b) = \mathrm{P}(a \mid b)$ [IDEMPOTENCE] *Q. E. D.*

THEOREM A.94. $\mathrm{P}(((aa)b)c \mid d) = \mathrm{P}(a(ab) \mid cd)\,\mathrm{P}(c \mid d) = \mathrm{P}(((ab)a)c \mid d)$

Proof:

(Axiom III.2) $\mathrm{P}(ab \mid c) = \mathrm{P}(a \mid bc)\,\mathrm{P}(b \mid c)$ [MULTIPLICATION]

(Thm. A.62) $\mathrm{P}((ab)c \mid d) = \mathrm{P}(a(bc) \mid d)$
[ASSOCIATION FOR FIRST ARGUMENT]

(Thm. A.40) $\mathrm{P}(ab \mid c) = \mathrm{P}(ba \mid c)$
[COMMUTATION OF FIRST ARGUMENT] *Q. E. D.*

THEOREM A.95. **[Distribution Lemma]**
$\mathbf{P(a(bc) \mid d) = P((ab)(ac) \mid d)}$

Proof:

(Thm. A.93) $\mathrm{P}(a(bc) \mid d) = \mathrm{P}(a \mid (bc)d)\,\mathrm{P}(bc \mid d) = \mathrm{P}((aa)(bc) \mid d)$

(Thm. A.94) $\mathrm{P}(((aa)b)c \mid d) = \mathrm{P}(a(ab) \mid cd)\,\mathrm{P}(c \mid d) = \mathrm{P}(((ab)a)c \mid d)$

(Thm. A.62) $\mathrm{P}((ab)c \mid d) = \mathrm{P}(a(bc) \mid d)$
[ASSOCIATION FOR FIRST ARGUMENT] *Q. E. D.*

THEOREM A.96. $\mathrm{P}(\neg(\neg(ab)\neg(ac)) \mid d) = \mathrm{P}(ab \mid d) + \mathrm{P}(ac \mid d) - \mathrm{P}((ab)(ac) \mid d)$

Proof:

(Thm. A.90) $\mathrm{P}(\neg(\neg a\neg b) \mid c) = \mathrm{P}(a \mid c) + \mathrm{P}(b \mid c) - \mathrm{P}(ab \mid c)$ *Q. E. D.*

THEOREM A.97. **[Distribution of Boolean Sum]**
$\mathbf{P(a\neg(\neg b\neg c) \mid d) = P(\neg(\neg(ab)\neg(ac)) \mid d)}$

Proof:

(Thm. A.92) $\mathrm{P}(a\neg(\neg b\neg c) \mid d) = \mathrm{P}(ab \mid c) + \mathrm{P}(ac \mid d) - \mathrm{P}(a(bc) \mid d)$

(Thm. A.96) $\mathrm{P}(\neg(\neg(ab)\neg(ac)) \mid d) = \mathrm{P}(ab \mid d) + \mathrm{P}(ac \mid d) - \mathrm{P}((ab)(ac) \mid d)$

(Thm. A.95) $\mathrm{P}(a(bc) \mid d) = \mathrm{P}((ab)(ac) \mid d)$ [DISTRIBUTION LEMMA] *Q. E. D.*

THEOREM A.98. $\mathrm{P}(\neg(\neg b(\neg\neg b))a \mid c) = \mathrm{P}(\neg(\neg b(\neg\neg b)) \mid ac)\,\mathrm{P}(a \mid c) = \mathrm{P}(a \mid c)$

Proof:

(Axiom III.2) $\mathrm{P}(ab \mid c) = \mathrm{P}(a \mid bc)\,\mathrm{P}(b \mid c)$ [MULTIPLICATION]

(Thm. A.82) $\mathrm{P}(\neg(\neg aa) \mid b) = 1 = \mathrm{P}(\neg(a\neg a) \mid b)$ *Q. E. D.*

THEOREM A.99. $\mathrm{P}(\neg(\neg(ab)\neg(ab)) \mid c) = \mathrm{P}(a \mid c)$

Proof:

(Thm. A.97) $\mathrm{P}(a\neg(\neg b\neg c) \mid d) = \mathrm{P}(\neg(\neg(ab)\neg(ac)) \mid d)$

(Thm. A.98) $\mathrm{P}(\neg(\neg b(\neg\neg b))a \mid c) = \mathrm{P}(\neg(\neg b(\neg\neg b)) \mid ac)\,\mathrm{P}(a \mid c) = \mathrm{P}(a \mid c)$

(Thm. A.40) $\mathrm{P}(ab \mid c) = \mathrm{P}(ba \mid c)$
[COMMUTATION OF FIRST ARGUMENT] *Q. E. D.*

THEOREM A.100. $\mathrm{P}((\neg\neg a)b \mid c) = \mathrm{P}(ab \mid c)$

Proof:

(Thm. A.68) $\mathrm{P}(ab \mid c) + \mathrm{P}(\neg ab \mid c) = \mathrm{P}(cb \mid c) + \mathrm{P}(\neg cb \mid c)$

(Axiom III.2) $\mathrm{P}(ab \mid c) = \mathrm{P}(a \mid bc)\,\mathrm{P}(b \mid c)$ [MULTIPLICATION] *Q. E. D.*

THEOREM A.101. $\mathrm{P}(a \mid c) = \mathrm{P}(b \mid c) \Rightarrow \mathrm{P}(\neg a \mid c) = \mathrm{P}(\neg b \mid c)$

Proof:

(Thm. A.64) $\mathrm{P}(a \mid b) + \mathrm{P}(\neg a \mid b) = 1 + \mathrm{P}(\neg b \mid b)$ *Q. E. D.*

THEOREM A.102. $\mathrm{P}((\neg\neg(\neg a\neg b))\neg c \mid d) = \mathrm{P}(\neg a(\neg\neg(\neg b\neg c)) \mid d)$

Proof:

(Thm. A.62) $\mathrm{P}((ab)c \mid d) = \mathrm{P}(a(bc) \mid d)$
[ASSOCIATION FOR FIRST ARGUMENT]

(Thm. A.100) $\mathrm{P}((\neg\neg a)b \mid c) = \mathrm{P}(ab \mid c)$

(Thm. A.40) $\mathrm{P}(ab \mid c) = \mathrm{P}(ba \mid c)$
[COMMUTATION OF FIRST ARGUMENT] *Q. E. D.*

THEOREM A.103. **[Association for Boolean Sum]**
$\mathbf{P}(\neg((\neg\neg(\neg a\neg b))\neg c) \mid d) = \mathbf{P}(\neg(\neg a(\neg\neg(\neg b\neg c))) \mid d)$

Proof:

(Thm. A.101) $\mathrm{P}(a \mid c) = \mathrm{P}(b \mid c) \Rightarrow \mathrm{P}(\neg a \mid c) = \mathrm{P}(\neg b \mid c)$

(Thm. A.102) $\mathrm{P}((\neg\neg(\neg a\neg b))\neg c \mid d) = \mathrm{P}(\neg a(\neg\neg(\neg b\neg c)) \mid d)$ *Q. E. D.*

THEOREM A.104. **[Commutation for Boolean Sum]**
$\mathbf{P}(\neg(\neg a\neg b) \mid c) = \mathbf{P}(\neg(\neg b\neg a) \mid c)$

Proof:

(Thm. A.40) $\mathrm{P}(ab \mid c) = \mathrm{P}(ba \mid c)$
[COMMUTATION OF FIRST ARGUMENT]

(Thm. A.101) $\mathrm{P}(a \mid c) = \mathrm{P}(b \mid c) \Rightarrow \mathrm{P}(\neg a \mid c) = \mathrm{P}(\neg b \mid c)$ *Q. E. D.*

THEOREM A.105. **[Idempotence for Boolean Sum]**
$\mathbf{P}(\neg(\neg a\neg a) \mid b) = \mathbf{P}(a \mid b)$

Proof:

(Thm. A.30) $\mathrm{P}(aa \mid b) = \mathrm{P}(a \mid b)$ [IDEMPOTENCE]

(Thm. A.100) $\mathrm{P}((\neg\neg a)b \mid c) = \mathrm{P}(ab \mid c)$

(Thm. A.101) $\mathrm{P}(a \mid c) = \mathrm{P}(b \mid c) \Rightarrow \mathrm{P}(\neg a \mid c) = \mathrm{P}(\neg b \mid c)$ *Q. E. D.*

THEOREM A.106. $\mathrm{P}(a \mid b) = \mathrm{P}(a \mid b\neg(c\neg c))$

Proof:

(Thm. A.98) $\mathrm{P}(\neg(\neg b(\neg\neg b))a \mid c) = \mathrm{P}(\neg(\neg b(\neg\neg b)) \mid ac)\,\mathrm{P}(a \mid c) = \mathrm{P}(a \mid c)$

(Thm. A.40) $\mathrm{P}(ab \mid c) = \mathrm{P}(ba \mid c)$
[COMMUTATION OF FIRST ARGUMENT]

(Axiom II.2) $\forall c\,\mathrm{P}(a \mid c) = \mathrm{P}(b \mid c) \Rightarrow \mathrm{P}(d \mid a) = \mathrm{P}(d \mid b)$
[SUBSTITUTIVITY] *Q. E. D.*

THEOREM A.107. $\mathrm{P}(a \mid b)\,\mathrm{P}(b) = \mathrm{P}(ab)$

Proof:

(Thm. A.106) $\mathrm{P}(a \mid b) = \mathrm{P}(a \mid b\neg(c\neg c))$

(Axiom III.2) $\mathrm{P}(ab \mid c) = \mathrm{P}(a \mid bc)\,\mathrm{P}(b \mid c)$ [MULTIPLICATION]

(Thm. A.83) $\mathrm{P}(a) = \mathrm{P}(a \mid \neg(a\neg a)) = \mathrm{P}(a \mid \neg(\neg aa)) = \mathrm{P}(a \mid \neg(b\neg b)) = \mathrm{P}(a \mid \neg(\neg bb))$ *Q. E. D.*

THEOREM A.108. **[Conventional definition of relative probability]**
$\mathbf{P}(b) \neq \mathbf{0} \Rightarrow \mathbf{P}(a \mid b) = \mathbf{P}(ab)/\mathbf{P}(b)$

Proof:

(Thm. A.107) $\mathrm{P}(a \mid b)\,\mathrm{P}(b) = \mathrm{P}(ab)$ *Q. E. D.*

This shows that Popper's axioms exactly mimic the ordinary ones in the circumstance that $\mathrm{P}\,(b)$ is not zero. Therefore there is no clash between the two axiomatizations of probability. There is, however, a clash between the ordinary axiomatization and first order logic. To show this, in the next section, we show that the assignment of probabilities automatically produces a structure that is a Boolean algebra. In the case of statements, this produces the usual structure of *and, or, not, implication,* and *logical equivalence* with *true* implying probability one and *false* implying probability zero.

THEOREM A.109. $\forall x\, \mathrm{P}(a \mid x) = \mathrm{P}(b \mid x) \Rightarrow \mathrm{P}(ay \mid z) = \mathrm{P}(by \mid z)$

Proof:

(Axiom III.2) $\mathrm{P}(ab \mid c) = \mathrm{P}(a \mid bc)\,\mathrm{P}(b \mid c)$ [MULTIPLICATION] *Q. E. D.*

THEOREM A.110. $\forall x\, \mathrm{P}(a \mid x) = \mathrm{P}(b \mid x) \Rightarrow \mathrm{P}(y \mid az) = \mathrm{P}(y \mid bz)$

Proof:

(Thm. A.109) $\forall x\, \mathrm{P}(a \mid x) = \mathrm{P}(b \mid x) \Rightarrow \mathrm{P}(ay \mid z) = \mathrm{P}(by \mid z)$

(Axiom II.2) $\forall c\, \mathrm{P}(a \mid c) = \mathrm{P}(b \mid c) \Rightarrow \mathrm{P}(d \mid a) = \mathrm{P}(d \mid b)$
[SUBSTITUTIVITY] *Q. E. D.*

THEOREM A.111. $\forall x\, ((\mathrm{P}(a \mid x) = \mathrm{P}(b \mid x))\,\&\,(\mathrm{P}(c \mid x) = \mathrm{P}(d \mid x))) \Rightarrow$
$\mathrm{P}(ac \mid y) = \mathrm{P}(bd \mid y)$

Proof:

(Thm. A.110) $\forall x\, \mathrm{P}(a \mid x) = \mathrm{P}(b \mid x) \Rightarrow \mathrm{P}(y \mid az) = \mathrm{P}(y \mid bz)$

(Axiom III.2) $\mathrm{P}(ab \mid c) = \mathrm{P}(a \mid bc)\,\mathrm{P}(b \mid c)$ [MULTIPLICATION] *Q. E. D.*

THEOREM A.112. **[Substitutional Equivalence]**
$\forall \boldsymbol{x}\, \mathbf{P}(\boldsymbol{a} \mid \boldsymbol{x}) = \mathbf{P}(\boldsymbol{b} \mid \boldsymbol{x}) \Rightarrow$
a **may be substituted for some or all occurences of *b* in any well-formed formula without changing it truth value.**

Proof:

(Axiom II.2) $\forall c\, \mathrm{P}(a \mid c) = \mathrm{P}(b \mid c) \Rightarrow \mathrm{P}(d \mid a) = \mathrm{P}(d \mid b)$
[SUBSTITUTIVITY] (Substituting the conditional)

(Thm. A.101) $\mathrm{P}(a \mid c) = \mathrm{P}(b \mid c) \Rightarrow \mathrm{P}(\neg a \mid c) = \mathrm{P}(\neg b \mid c)$ (Substituting the complement)

(Thm. A.111) $\forall x\, ((\mathrm{P}(a \mid x) = \mathrm{P}(b \mid x))\,\&\,(\mathrm{P}(c \mid x) = \mathrm{P}(d \mid x))) \Rightarrow$
$\mathrm{P}(ac \mid y) = \mathrm{P}(bd \mid y)$ (Substituting the conjunction) *Q. E. D.*

THEOREM A.113. If a and $b \in S$, then $a \vee b \in S$

Proof:

(Axiom III.1) $\mathrm{P}(ab \mid c) \leq \mathrm{P}(a \mid c)$ [MONOTONY]

(Axiom III.2) $\mathrm{P}(ab \mid c) = \mathrm{P}(a \mid bc)\,\mathrm{P}(b \mid c)$ [MULTIPLICATION]

(Def. A.2) $a \equiv b \vee c \Leftrightarrow a \equiv \neg(\neg b \neg c)$

(Def. A.1) $a \equiv b \Leftrightarrow \forall x\, \mathrm{P}(a \mid x) = \mathrm{P}(b \mid x)$

(Thm. A.101) $\mathrm{P}(a \mid c) = \mathrm{P}(b \mid c) \Rightarrow \mathrm{P}(\neg a \mid c) = \mathrm{P}(\neg b \mid c)$

(Thm. A.111) $\forall x\, ((\mathrm{P}(a \mid x) = \mathrm{P}(b \mid x))\, \&\, (\mathrm{P}(c \mid x) = \mathrm{P}(d \mid x))) \Rightarrow$
$\mathrm{P}(ac \mid y) = \mathrm{P}(bd \mid y)$ *Q. E. D.*

THEOREM A.114. If $a \in S$, then $\neg a \in S$

Proof:

(Axiom IV.1) $\exists b\, \mathrm{P}(a \mid a) \neq \mathrm{P}(b \mid a) \Rightarrow \mathrm{P}(a \mid a) = \mathrm{P}(c \mid a) + \mathrm{P}(\neg c \mid a)$
[COMPLEMENTATION] *Q. E. D.*

THEOREM A.115. $a \vee b \equiv b \vee a$

Proof:

(Thm. A.104) $\mathrm{P}(\neg(\neg a \neg b) \mid c) = \mathrm{P}(\neg(\neg b \neg a) \mid c)$

(Def. A.2) $a \equiv b \vee c \Leftrightarrow a \equiv \neg(\neg b \neg c)$ *Q. E. D.*

THEOREM A.116. $(a \vee b) \vee c \equiv a \vee (b \vee c)$

Proof:

(Thm. A.103) $\mathrm{P}(\neg((\neg\neg(\neg a \neg b))\neg c) \mid d) = \mathrm{P}(\neg(\neg a(\neg\neg(\neg b \neg c))) \mid d)$
[ASSOCIATION FOR BOOLEAN SUM]

(Def. A.2) $a \equiv b \vee c \Leftrightarrow a \equiv \neg(\neg b \neg c)$ *Q. E. D.*

THEOREM A.117. $a \vee a \equiv a$

Proof:

(Thm. A.105) $\mathrm{P}(\neg(\neg a \neg a) \mid b) = \mathrm{P}(a \mid b)$
[IDEMPOTENCE FOR BOOLEAN SUM]

(Def. A.2) $a \equiv b \vee c \Leftrightarrow a \equiv \neg(\neg b \neg c)$ *Q. E. D.*

THEOREM A.118. $ab \vee a\neg b \equiv a$

Proof:

(Thm. A.99) $\mathrm{P}(\neg(\neg(ab)\neg(ab)) \mid c) = \mathrm{P}(a \mid c)$

(Def. A.2) $a \equiv b \vee c \Leftrightarrow a \equiv \neg(\neg b \neg c)$ *Q. E. D.*

THEOREM A.119. $\exists a \exists b\, a \not\equiv b$

Proof:

(Thm. A.27) $\exists a\, \mathrm{P}(\neg a \mid a) = 0$

(Thm. A.82) $\mathrm{P}(\neg(\neg a a) \mid b) = 1 = \mathrm{P}(\neg(a \neg a) \mid b)$

(Thm. A.101) $\mathrm{P}(a \mid c) = \mathrm{P}(b \mid c) \Rightarrow \mathrm{P}(\neg a \mid c) = \mathrm{P}(\neg b \mid c)$

(Def. A.1) $a \equiv b \Leftrightarrow \forall x\, \mathrm{P}(a \mid x) = \mathrm{P}(b \mid x)$ *Q. E. D.*

Definition A.3. **[Entailment]** $\boldsymbol{a \supseteq b \Leftrightarrow ab \equiv b}$

Theorem A.120. $\boldsymbol{a \supseteq b \Rightarrow \mathrm{P}(a \mid b) = 1}$

But it is not the case that $\mathrm{P}(a \mid b) = 1 \Rightarrow a \supseteq b$. Consider probability in the universe of positive integers. Let the probability be defined as:

$$\mathrm{P}(a \mid b) = \lim_{n\to\infty} \frac{\{\text{number of integers } \in a\} \leq n}{\{\text{number of integers } \in b\} \leq n}$$

The case of prime numbers can illustrate How the converse of Theorem A.120 is false. Let $a = \{\text{all primes}\}$ and $b = \{\text{all positive integers}\}$, then $\mathrm{P}(\neg a \mid b) = 1$, but it is not the case that no number is a prime.

Theorem A.121. $a \supseteq b \Leftrightarrow \mathrm{P}(a \mid \neg ab) \neq 0$

Theorem A.122. $\boldsymbol{a \supseteq b \Leftrightarrow \mathrm{P}(a \mid \neg ab) = 1}$

Bayes' Theorem

Theorem A.123. **[Bayes' Theorem]**
$\boldsymbol{\mathrm{P}(a \mid bc)\,\mathrm{P}(b \mid c) = \mathrm{P}(b \mid ac)\,\mathrm{P}(a \mid c)}$

Proof:

(Axiom III.2)	$\mathrm{P}(ab \mid c) = \mathrm{P}(a \mid bc)\mathrm{P}(b \mid c)$ [Multiplication]
(Thm. A.40)	$\mathrm{P}(ab \mid c) = \mathrm{P}(ba \mid c)$
	[Commutation of First Argument] *Q. E. D.*

Theorem A.124. $\exists k\,\forall a \quad k\,\mathrm{P}(a \mid bc) = \mathrm{P}(b \mid ac)\mathrm{P}(a \mid c)$

Proof:

Thm. A.123	$\mathrm{P}(a \mid bc)\mathrm{P}(b \mid c) = \mathrm{P}(b \mid ac)\mathrm{P}(a \mid c)$ [Bayes' Theorem]
	Let $k = \mathrm{P}(b \mid c)$ *Q. E. D.*

This is an alternative rendering of Bayes' Theorem usually read as: *posterior probability is proportional to likelihood × prior probability.*

Theorem A.125. If none of the denominator components $= 0$, then $\forall d \forall e \;\; \frac{\mathrm{P}(d|bc)}{\mathrm{P}(e|bc)} = \frac{\mathrm{P}(b|dc)}{\mathrm{P}(b|ec)}\frac{\mathrm{P}(d|c)}{\mathrm{P}(e|c)}$

Proof:

Thm. A.123	$\mathrm{P}(a \mid bc)\mathrm{P}(b \mid c) = \mathrm{P}(b \mid ac)\mathrm{P}(a \mid c)$ [Bayes' Theorem]
	Use Thm. A.123 twice, once with $a = d$ and once with $a = e$
	Divide equals by and cancel $\mathrm{P}(b \mid c)$ *Q. E. D.*

This version of Bayes' Theorem is usually rendered as: *posterior odds = liklihood ratio × prior odds.*

The following two theorems cannot be laid at the feet of Karl Popper although I suspect he would be amused by the sharp stick in the eye of the Bayesian community.

Semantic Expansion

THEOREM A.126. **[Expansion of S] If $x \not\subseteq S$ and $a, b, c \in S$ and $x \Leftrightarrow a$, then there exists a probability space, $\mathrm{P}'(\cdot \mid \cdot)$ on $S' = S \cup x$.**

Proof:

$\forall b \forall c$ Let $\mathrm{P}'(xc \mid b) = \mathrm{P}(ac \mid b)$ and

Let $\mathrm{P}'(b \mid xc) = \mathrm{P}(b \mid ac)$ *Q. E. D.*

The immediate use of this proposition is to extend the probability of an event to the announcement of the event. It is an unusual property. Consider an alternative. Let a be a pot of boiling water at sea level. Let x be the an observer saying, "The pot of water is boiling at sea level." Assume $x \Leftrightarrow a$. It mapes sense to say, "$temperature(a) = 212°\mathrm{F}$," but it does not make sense to say, "$temperature(x) = 212°\mathrm{F}$." It is simply inexplicable usage to say, "The temperature of the observer saying, 'The water is boiling,' is 212°F."

On the other hand, if a is snake eyes coming up when two dice are thrown, b, and x is an observer yelling, "Snake eyes" (assuming he does not lie); then if $\mathrm{P}(a \mid b) = \frac{1}{36}$, it also makes sense to assert, $\mathrm{P}(\text{"Snake eyes"} \mid b) = \frac{1}{36}$.

Joining Multiple Probabilities

THEOREM A.127. **[Splicing] Let $\{\mathrm{P}_{h_1}(\cdot \mid \cdot), \ldots, \mathrm{P}_{h_n}(\cdot \mid \cdot)\}$ be n probability spaces on S and $P = \{p_1, \ldots, p_n\}$, n real numbers, such that, $\Sigma_{i=1}^{n} p_i = 1$ and $p_i \geq 0$. In addition, let $U = \{u_1, \ldots, u_n\}$ be a set of n mutually exclusive elements such that, $u_i \not\subseteq S$. If this is so, then there exists a probability distribution, $\mathrm{P}_\Sigma(\cdot \mid \cdot)$ on $S \cup U$.**

Proof:

Let $a, b, c \in S$ and $x, y, z \in S \cup U$

Assign probabilities in this way:

i Let $\mathrm{P}_\Sigma(u_i \mid \neg(x\neg x)) = \mathrm{P}_\Sigma(u_i) = p_i$

ii $\mathrm{P}_\Sigma(a \mid b) = p_1\mathrm{P}_1(a \mid b) + p_2\mathrm{P}_2(a \mid b) + \cdots + p_n\mathrm{P}_n(a \mid b)$

iii $\mathrm{P}_\Sigma(au_i \mid b) = p_i\mathrm{P}_i(a \mid b)$

iv $\mathrm{P}_\Sigma(a\neg u_i \mid b) = \mathrm{P}_\Sigma(a \mid b) - \mathrm{P}_\Sigma(au_i \mid b) = \mathrm{P}_\Sigma(a \mid b) - p_i\mathrm{P}_\Sigma(a \mid b)$

v $\mathrm{P}_\Sigma(a \mid bu_i) = \mathrm{P}_i(a \mid b)$

vi $\mathrm{P}_\Sigma(u_i \mid u_i) = 1$

vii $\mathrm{P}_\Sigma(x \mid u_iu_jy) = 1$

viii $\mathrm{P}_\Sigma(xu_i \mid bu_j) = 0$

ix $\mathrm{P}_\Sigma(u_iu_jx) = 0$

x $\mathrm{P}_\Sigma(xy \mid z) = \mathrm{P}_\Sigma(x \mid yz)\,\mathrm{P}_\Sigma(y \mid z)$

xi $\mathrm{P}_\Sigma(x \mid yz)\,\mathrm{P}_\Sigma(y \mid z) = \mathrm{P}_\Sigma(y \mid xz)\mathrm{P}_\Sigma(x \mid y)$ Given these assignments it is easy to show that the combined system follows the probability axioms. *Q. E. D.*

Thus it is the case that Bayes' Theorem:

$$\mathrm{P}_\Sigma(u_i \mid a) - \mathrm{P}_\Sigma(u_i)\frac{\mathrm{P}_\Sigma(a \mid u_i)}{\mathrm{P}_\Sigma(a)}$$

is true.

It holds for $\mathrm{P}_\Sigma(\cdot \mid \cdot)$ no matter what is meant by P and U. U could be {Wynken, Blynken, Nod} or {Snap, Crackle, Pop} and P could be the odds of {Win, Place, Show} in the last Kentucky Derby divided by their sum or any an other three fractions that sum to one.

The threshold for syntactic consistency is vanishingly small; the threshold for meaning is more problematic. If and only if p_i bears the same relationship to u_i as $\mathrm{P}_{h_j}(a \mid b)$ does to $\{a, b\} \in S$, is there no problem.

When the n hypotheses, $h_1 \equiv \mathrm{P}_{h_1}(\cdot \mid \cdot)$, $h_2 \equiv \mathrm{P}_{h_2}(\cdot \mid \cdot)$, $\ldots$, $h_n \equiv \mathrm{P}_{h_n}(\cdot \mid \cdot)$ take the place of $\{u_i, u_2, \ldots, u_n\}$ and are cobbled together by $\{p_1, \ldots, p_n\}$, arbitrary numbers in the guise of subjective probability, a semantic train wreck occurs.

It is almost always the case that $\mathrm{P}_{h_i}(\cdot \mid \cdot)$ refers to propensities or relative frequencies or some other physical or situational property of $\{a, b, \ldots\} \in S$, not a subjective feeling or a bias-neutral straw man. If the choice of h_i actually were determined by a chance event, the combination would be semantically meaningful. It must be that God actually *does* play dice with the universe, not only pertaining to very small things, but also in deciding what rules the universe will follow. Otherwise, the combination is syntactically correct, but meaningless.

If there is not semantic rigor,

$$\mathrm{P}_\Sigma(h_i \mid e) = \mathrm{P}_\Sigma(h_i)\frac{\mathrm{P}_\Sigma(e \mid x_i)}{\mathrm{P}_\Sigma(e)}$$

does not produce the probability of the truth of hypothesis, h_i, given evidence, e. It just produces a meaningless number that is consistent with the axioms of probability. It is the result of an invalid attempt to transform likelihoods into probabilities of truth.

Made in the USA
Monee, IL
02 November 2026

67376111R00109